# THE
# ORIGIN
## OF
# MINDS

# THE
# ORIGIN
## OF
# MINDS

*Evolution, Uniqueness,*
*and the New Science of the Self*

Peggy La Cerra & Roger Bingham

Harmony Books, New York

Published by Harmony Books, New York, New York.
Member of the Crown Publishing Group, a division of
Random House, Inc.

www.randomhouse.com

HARMONY BOOKS is a registered trademark and the
Harmony Books colophon is a trademark of Random
House, Inc.

The authors and publisher gratefully acknowledge permission to reprint from the poem "In Memory of W. B.
Yeats," from *Collected Poems* by W. H. Auden. Repro-
duced by permission of Faber & Faber Ltd.

Printed in the United States of America

DESIGN BY ELINA D. NUDELMAN

Library of Congress Cataloging-in-Publication Data

La Cerra, Peggy.
The origin of minds : evolution, uniqueness, and the
new science of the self / Peggy La Cerra & Roger
Bingham.
p. cm.
1. Brain–Evolution. 2. Genetic psychology. 3. Human
evolution. I. Bingham, Roger. II. Title.
QP376 .L25 2002
155.7–dc21
2002005707

ISBN 0-609-60558-5

10  9  8  7  6  5  4  3  2  1

First Edition

*To my family:*
*Gae, Fred, Marge, Michael, and Laurie La Cerra,*
*Lisa and Justin Nelson,*
*and*
*Behzad Boroumand*
**–PLC**

*To the memory of my father, Arthur Bingham,*
*and in celebration of the women in my life:*
*my mother, Flo; my wife, Linda; my sister, Jeanne;*
*and my mother-in-law, Ronnie.*
**–RB**

# CONTENTS

# Contents

# INTRODUCTION
*A Day in the Life*

On any given day, 400,000 children are born. From the suburbs of Chicago to the shantytowns of Calcutta, from the deserts of Africa to the rain forests of South America, you can hear the first cries of these new players in the markets of life—each one the voice of a unique individual.

And on any given day, we encounter the consequences of this individuality. On June 30, 1905, an obscure clerk in the Swiss Patent Office publishes a paper about relativity that alters our understanding of the physical world. On December 1, 1955, a small black woman refuses to go to the back of a bus in Montgomery, Alabama, and galvanizes a racial revolution. And on September 11, 2001, a Saudi Arabian exiled in Afghanistan ignites a terrorist attack that will change the life of every American.

On any given day, individuals make history. But how are individuals made? What mechanism explains the creation of six billion different minds—minds capable of thinking and behaving in radically different ways? Where could you find a scientifically rigorous theory that was at once universally applicable, yet generated the astonishing uniqueness–the individuality–that we know from everyday personal experience to be a fact of life?

Let's begin with *The Origin of Species*. Published in 1859, Charles Darwin's book was a scientific Genesis. In his theory of natural selection, Darwin explained how changing environmental circumstances acted as a kind of sieve, winnowing out some life forms and preserving others, which then went on to reproduce and create a niche for themselves. We owe our very existence to these unpredictable challenges to the temporary order of things. Strong evidence suggests that an asteroid impact 65 million years ago proved fatal to the dinosaur dynasty that had ruled for 140 million years. In a geologically brief moment of time, tiny mammals–until then bit players in the grand pageant of life–found themselves thrust onto center stage. And from so simple a beginning, we humans eventually evolved.

Darwin's story offers a compelling explanation for the different anatomical and physiological variations of life forms. Take one example–the beaks of birds. Birds evolved in a world of seeds to be broached, or insects to be caught on the wing, larvae to be drilled for in tree barks, mice to be swooped down upon, mollusks to be dashed on a rock, or fish to be scooped from the oceans. Birds have evolved the specialized tools–the adaptations–that have allowed them to solve the problem of surviving and reproducing in these different environments, and they have passed on the genetic recipe for making those tools to their offspring.

Look at the beak of a raptor such as a hawk, the pouch of a pelican, the pneumatic drill of a woodpecker, and you

✓ INTERNAL MUTATIONS    [2]
∿ EXTERNAL MUTATIONS

have an example of an adaptation. Specialized physical organs like livers dedicated to removing toxins, or hearts dedicated to pumping blood, are also adaptations. So is the opposable thumb that our primate ancestors found advantageous in grasping food and that you may now find advantageous in turning the pages of this book.

You might argue that, if this evolutionary logic works for beaks and hearts, then perhaps it should work for brains and the minds they generate. This is the claim made by evolutionary psychologists who suggest that the mind is a collection of specialized and—here's the key point—*heritable* mental organs designed by natural selection to solve adaptive problems, like finding food or choosing a mate.

The assumption is that natural selection would have shaped an intelligence system in just the same way that it would shape a bodily organ. But the brain is not a beak; the intelligence system is not a liver or an opposable thumb. The evolutionary psychologists' leap of logic makes sense only if the environmental challenges remain static enough to sculpt an instinct over evolutionary time—to etch the instructions for its reliable development into your genome.

But what happens when the environment is *dynamic*—like the social world? What happens when the problem to be solved—like choosing a mate or deciding upon a coalition partner—becomes a moving target, changing not over evolutionary time, nor even within the life span of the individual—but moment by moment?

Ask yourself: What kind of mind could be responsive to real-time changes in the social world? What kind of mind can generate a Self—an internal agent that knows what behavior might be effective at any given time and can navigate you through the social landscape, altering in the moment, from situation to situation? What kind of mind can keep track of your past and present interactions in this changing human social market and generate possible futures—creative acts of imagination?

In September 1998, we published a blueprint for the human intelligence system—a new evolutionary model of the mind—in the *Proceedings of the National Academy of Sciences.* We suggested that the mind arises from an evolved intelligence system that was designed to be responsive to any number of adaptive problems (including those yet to emerge in the evolutionary future). But the *fundamental* problem it was designed to solve—the problem that drove the elemental logic inherent in all intelligence systems on earth—was one imposed by the universe at large, by the physical laws of energy.

Here's the crux of the argument. To do *anything*—locate food, find a mate, reproduce, compose a sonata, solve an equation—you have to stay alive with enough surplus energy to perform the task at hand. Energy management drove the foundational adaptive design of all ancestral intelligence systems. And all subsequent design features evolved as integrated augmentations of this core system—including the part that ultimately gives rise to your mind. This remarkable adaptation is designed to deal with the problem of behavioral energy management in an ever-fluctuating environment *by changing itself with every experience.* And this design feature underlies all of our most impressive abilities as humans.

This book presents a novel framework for understanding the origin—and nature—of minds, and the genesis of all intelligence systems. It connects the intelligence systems of bacteria—the opening act in the four-billion-year drama of life on this planet—to the intelligence systems of human beings today. It validates individual uniqueness as an adaptive expression of human nature, while acknowledging the influence of the social world in the creation of our sense of self. And it makes a bold new claim: that we are—all of us—unique *by design.*

The key component of this framework is what we call an adaptive representational network (ARN). Think of it as a network of neurons that memorializes a brief scene in the ongoing movie of your life, linking together your physical and emotional state, the environment you are in, the behavior or thought you generate, and the problem-solving outcome. We think of ARNs as the fundamental units of intelligence—rather like atoms of behavior. Here's a rough analogy: If you take two atoms of hydrogen and an atom of oxygen something quite extraordinary results. You can drink, or go swimming in, what emerges from the union. In similar fashion, the combinatorial properties of ARNs lead to a surprising range of human abilities—the creation of selves and personalities, the generation of unprecedented thoughts and metaphors, and the ability to make inferences about our world and the people with whom we share it—even the basis for our capacity to acquire and produce language.

*The Origin of Minds* is a book about evolution, uniqueness, and this new scientific model of the self. It takes you on a journey from the physical laws of the universe to the making of your mind. A journey that begins in your own garden.

# 1

## THE LAWS OF ENERGY AND THE DESIGN OF YOUR MIND

Imagine that you are lazing in a garden on a summer afternoon, vaguely aware of the insistent droning of bees and the occasional high-performance aerobatics of a hummingbird. Butterflies are wafting gently along on their tipsy trajectories, drifting down to perfect landings on a gloriously colored array of flowers. Everything seems tranquil, leisurely.

And yet, in every corner of this landscape, deals are being struck. There are buyers and sellers, advertisers and consumers. This is a biological marketplace. And what you are witnessing–in your own backyard–is the operation of fundamental principles that explain the design of your mind.

This may seem a surprising claim. What could something so sophisticated as the human mind have in common with bees foraging and flowers being fertilized? The answer is: everything. All the neural machinery that supports your highest cognitive abilities (like memory, decision-making, modeling potential futures, imagining this garden) is a deluxe, augmented version of an evolutionarily ancient system to be found in bees, hummingbirds, butterflies—even bacteria. And the foundational requirement of the system is keeping an individual organism's energy account in the black.

The Bank of Nature does not tolerate overdrafts. The bee you are watching in your mind's eye is not meandering aimlessly, out for a joyride; it's on a mission—locating a source of energy to replace the fuel it is burning, and delivering a crucial surplus to the hive. Its navigational skills, memory, and sensory systems have evolved precisely to improve its ability to get the goods it needs for survival, and for the leveraging of its genes into the future.

Your intelligence system evolved to solve the same basic problems, and, as with the bee's system, the first problem to be solved is meeting the bioenergetic bottom line. All other life problems are secondary. And all other aspects of intelligence—including, ultimately, the most subjective aspects of your mind—are shaped by this universal mandate of life. Your thoughts, your feelings, your emotional attachments to others, and, ultimately, your behavior in the social world are guided by an enormously complex evolved system that can do many things; but its grand architecture arises from a foundational logic that acquires, manages, and directs energy to get the work of life done in an efficient way. To elaborate on these claims, to understand why energy management is so important, and how your intelligence system evolved, you need to fire up a few billion more of those deluxe brain cells and imagine yourself back in the garden.

Let's start with the obvious. In the garden, there are plants and animals. The plants are—well—planted: They don't move about in search of food. Compared to animals, plants have a limited repertoire. Plants are *autotrophs*, which means self-nourishing. Through the process of photosynthesis, the chlorophyll in plants uses the energy from sunlight to build carbohydrates (sugars) from carbon dioxide and water, liberating oxygen. The sugars are stored energy, stocking the plants' pantries.

Animals are *heterotrophs*, which means they are dependent on other organisms for food. Plants are the foundational resource for the survival of animals. Animals eat plants; some animals eat animals that have eaten plants. Carnivores or vegetarians—ultimately, we all dine off sunbeams.

For most of our evolutionary history, of course, packets of energy (food) were not handily available in some convenient one-stop supermarket, shrink-wrapped with nutritional information. In a state of nature, no delivery trucks respond instantly to consumer complaints of shortages; no one regularly restocks the shelves. Bees have to find flowers; changes in climate can radically change nectar concentrations. Berries and roots have to be located; lunch may have wings or legs and be inclined to escape.

And so, because resources are distributed, variable, and unpredictable, animals must forage to survive. They must move—which is energetically expensive. (If the hummingbird doesn't get a food fix within about thirty minutes, it has no fuel for takeoff.) The distribution and variability of resources means there are choices to evaluate and comparisons to be made—crucially, a memory of where resources are located. (Were those blue flowers a good source of nectar? Where did I cache those nuts?)

Take a bee's-eye view for a moment. The bee's primary mission is to collect nectar and pollen for food, and return it to the hive to rear baby bees. Nectar is basically sugar water (with some small amounts of amino acids, proteins, vitamins, enzymes, and other health supplements). Bees use nectar to make honey. Pollen—flower sperm—has a high protein content, plus fats, starches, and vitamins.

The evolutionary deal that bees (and other pollinators) have made with flowers is this: You supply a rich diet (energy); we pass on your genetic information (your DNA) by pollination. The mechanism, in the case of bees, is simple: As the bee probes deep into the flower to extract nectar, it is dusted with pollen. When the bee visits the next flower on its foraging rounds, the pollen is wiped off and triggers fertilization. The bee is a kind of DNA delivery service, and it is paid handsomely for its efforts.

Bees and flowers have been playing this game for well over 100 million years. They have evolved exquisitely sophisticated partnerships. The riotous colors of a garden are a result of this biological bartering. Bees have color vision, an acute sense of smell, and a brain capable of performing remarkably complex computations about the quality of the goods on offer. Flowers, in turn, have evolved increasingly tempting advertisements to seduce bees into sampling their wares. The symmetry, color, and odor of flowers are essentially biological billboards designed by natural selection to lure pollinators. To the bee, these are indications of nectar quality—a measure of the life-sustaining energy that comes from pollinating behavior.

For bees—for all living things—there is one simple, unavoidable fact of life: The energy input/output equation has to stay positive. According to the laws of thermodynamics, living things seem at first blush to be defying physics. Everything in the universe is inexorably moving toward a state of increasing disorder and randomness (entropy). And yet living things–examples of a high degree of organization– continue to flourish. How? The answer is that life forms owe their temporary victory against entropy to an evolved capacity to acquire energy and use it to survive and reproduce during their lifetimes.

Flowers, hummingbirds, bees–and you–have the ability to manage bioenergetic resources. Plants have evolved an integrated complex of hormonal systems–a coordinated, but noncentralized, intelligence system–that serves these purposes. But animals require a more sophisticated *centralized* system that constantly monitors and assesses inputs and outputs of energy; a system that maintains an energy surplus, ensuring survival.

At the core of our intelligence system is a component responsible for homeostasis–basically, self-regulation of physiological factors that are critical for energy management. Homeostasis involves activities like keeping body temperature, water and electrolyte balance, oxygen levels, and so on within a window that sustains life. The most important part of the brain involved in homeostasis is the hypothalamus. A pea-sized structure situated roughly in the middle of your brain (on a line with the brow ridges above your eyes), the hypothalamus sits like a spider at the center of a web of a number of brain systems.

At a very crude level, homeostatic systems do something rather like the job a thermostat does in maintaining a constant temperature in your home. The thermocouple senses deviations from a set point and activates cooling or heating systems. The autopilot of an airplane works on the same principles. Strictly speaking, a plane is *off* course for most of its flight. Variations in wind speed and direction, air density, fluctuations in engine efficiency, and so on are continuously sensed and compensated for. There is no magic here—just intricately engineered control systems converting constant feedback into revised solutions of the temperature problem or the optimal direction problem.

We life forms have similar problem-solving systems (though much more complex). Take body temperature. To support life, it must be maintained in a narrow window. Thirty-seven degrees centigrade (98.6 degrees Fahrenheit) is normal. At 33 degrees, you lose consciousness; at 28 degrees, your muscles fail; at 44 degrees, the proteins in your body are irreversibly damaged. Your body has evolved mechanisms to cope with deviations from normal. If the temperature rises, you perspire. The word is precise and evocative: It means to "breathe through" your skin. A bucket brigade of blood cells transports the excess heat from within your body to capillaries on the surface of your skin, where the heat evaporates through pores. If you are cold, you shiver. The surface capillaries constrict ("goose bumps"), muscle groups are activated, and your metabolic rate increases to produce warming.

These responses are tried and tested evolutionary solutions to problems, incorporated into the basic wiring of an organism. They are reflexes, produced automatically in response to an environmental challenge. You get hot, you perspire. You get cold, you shiver. This is not negotiable. It is in the nature of reflexes that no learning is necessary: You don't need lessons in shivering.

Reflexes are an essential part of an organism's behavioral repertoire, but they are limited. For one thing, they are involuntary: They short-circuit conscious thought, goals, and intentions. There is no weighing of costs and benefits, no internal debate of neurons about the best course of action given a changing situation. Reflexes were not designed to cope with environmental subtleties—just black-and-white emergencies. A reflex is Mother Nature's response to a 911 call.

There are more complex programmed sequences of actions called instincts. Again, virtually all members of a species are able to perform them without instruction (although they may improve their performance after instruction). Think of the weaving of webs by spiders, for example—behaviors so precise and intricately choreographed that we accept them as part of a species' innate endowment: built-in routines designed to ensure their survival and reproduction. This is fine provided that the environment remains static: that the world in which these routines evolved doesn't change. But suppose something unexpected happens. What then?

Suppose, for example, that you are a digger wasp (*Sphex ichneumoneus*). When the time comes to lay your eggs, your routine is to build a burrow, assault some unsuspecting cricket (stinging it so it is paralyzed but still alive), then drag it to the edge of your burrow. You make a quick inspection of the burrow, then drag the cricket inside. Then you lay your eggs, seal the burrow, and fly away. Eventually, the eggs hatch, the grubs feed on the comatose pickled cricket, and *voilà!*—more little Sphexes. Now suppose that one day you drag the cricket to the edge of your burrow and then make the obligatory last-minute inspection. But when you emerge, the cricket has been moved a few inches. What do you do? Scientists have played this nasty trick on Sphexes forty straight times, and the result is always the same: You drag the cricket to the edge of the burrow, then make a final inspection; you drag the cricket to the edge of the burrow, then make a final inspection; you drag the cricket . . . and so on, like a needle stuck in the groove of an old, scratched record. The egg-laying behavioral routine of Sphexes works perfectly if the environment is stable. But Sphexishness can be an evolutionary liability in a world of change—the world we live in. So: How did natural selection solve the problem of negotiating the uncertainties of the biological marketplace?

There are some clues in the behavior of one of the simplest of organisms—the bacterium *Escherichia coli* that lives in our intestinal tract. *E. coli* is about as basic as life gets: a single cell with a genome consisting of a single molecule of DNA (human beings have roughly a million million cells). *E. coli* looks like a small cocktail sausage with about six flagellar filaments (like the lashes of a whip) trailing behind it. If you laid 10,000 *E. coli* end to end, they would measure the length of your thumb. When the flagella turn counterclockwise, they braid together and propel the bacterium at a top speed of about one thumblength an hour: It is said to "run." When they turn clockwise, the braid comes apart, and the flagella flail independently: The bacterium is in "tumble" mode.

Why this intestinal gymnastics? *E. coli* is trying to make a living, seeking out the energy it needs to survive and avoiding toxins that could kill it. If *E. coli* senses sugars like glucose, for example (a source of energy), or the presence of amino acids (the building blocks of proteins), it runs; if the trail goes cold, it tumbles, casting about for new sources of nutrients.

*E. coli* has more than a dozen different types of receptors (protein molecules) on its surface that sample the liquid environment in your intestinal tract, detecting the presence of key chemicals. (Is this a sugar? A toxic metal?) The chemicals bind to the receptors, which are then said to be "occupied." Like a hotel manager checking room occupancy, *E. coli* monitors the change of occupancy of the various sensory receptors over time. (How many welcome sugar guests, how many toxic guests, and so on.) The period for comparing chemical concentrations is brief: about four seconds. In a sense, this is the memory of the bacterium. It compares the events of the past second with the preceding three seconds, computes the change in the type of guests occupying its receptors, figures out whether life is getting better or worse, and runs or tumbles accordingly.

Strictly speaking, *E. coli* possesses neither a nervous system nor a brain. But it does have what could be described as a centralized intelligence system. At a rudimentary level, it does the kind of things that bigger, top-of-the-line brains can do. It can integrate information from sensory mechanisms that detect biologically salient features of the environment. It has a central decision-making circuitry that encodes and analyzes information about its past and present to chart a course into a well-chosen future. And it has the equipment—the behavioral effector systems—to execute the plan.

Back in our garden, a bee on a nectar mission is using fundamentally the same evolved operational system when it veers away from one flower patch in favor of another. It integrates information from its senses with stored data about the success or failure of its earlier foraging flights to produce a decision—avoid the yellow flowers, for example. The major difference between the bee and the smart sausage is sheer computing power. For *E. coli*, the slate of memory is wiped clean every four seconds. It cannot profit from experience. The bee, on the other hand, has evolved neural tissue that can store records of past experiences that guide future behavior.

And you, of course—lazing in the garden, contemplating the bees while *E. coli* runs and tumbles through the remnants of your lunch—*you* are blessed with a quite extraordinary capacity for making enduring associations between events and their outcomes. For you, memory is a deep and rich resource. Countless past events infuse your experience of the present moment with unique significance. When you access the memory of an event, you can recall the action, the players, and the relationships between the players. As we'll explain in a later chapter, these versatile neural networks also allow you to build up detailed profiles of the people you interact with and their behavior. An inappropriate gesture, the timely payment of a debt, a kindness here, a falsehood there—all this is stored as neural representations of your lifetime experiences: the autobiographical record of your development as a unique individual. This is the scaffolding of your self. You are not constrained by the four-second slate of the bacterial sausage or the limited pilot's logbook of the bee. You have a complex memory system that can serve you for life.

Why? Primarily because you have an enormous neocortex. This is the area of the brain above and behind your eyes—the convoluted tissue that looks like a walnut or the appropriately named brain coral (except that it is soft). Your neocortex has a wonderful quality called neuroplasticity: It can adapt on-line to experience. This is a fairly recent discovery and goes against the old conventional wisdom of how brains work (that neuronal circuitry is basically in place at birth and it's all downhill from there). We now know that the brain is capable of fairly impressive rewiring. And the "higher" the cortical area (more recent in evolutionary terms), the more plastic it seems to be.

For example, when a blind person learns Braille, the area of the neocortex that processes sensory input from the tip of the reader's index finger grows larger: There is more neuronal real estate dedicated to this new, tactile task (and it is in the visual cortex—an area that would normally be activated in sighted people). The same thing happens if you learn to play a musical instrument such as a piano: The cortical area that corresponds to (or "represents") the fingers of the hand traversing the keys enlarges. This effect is especially pronounced in the area of cortex devoted to the left hand in musicians like guitarists or violinists, who are adept at fingering strings. Amazingly, even imagining or visualizing playing produces similar changes in the cortex.

There's a caveat. The task has to be relevant or "salient": something that impacts upon the individual's fitness or well-being. Events that are not salient don't produce changes in neuroplasticity—even if they happen with the same frequency as salient inputs. (In the piano-playing experiments we've just described, of course, the subjects were "performing": They knew that the results were important to the researchers, and to themselves as professionals.) Which raises several interesting questions. How does an individual organism assess the salience of an activity–the importance of a life event to the bottom energetic line? What motivates it to change its behavior? How is that decision made?

The answers to these questions will soon become clear. For now, the important point is this: None of the issues we have touched on so far–from bioenergetic management of resources to reflexes and instincts, from the plastic neural substrate of the neocortex to the psychology of motivation– can sensibly be treated in isolation. They are all part of an integrated intelligence system, fashioned by natural selection and yoked together over evolutionary time.

We're presenting a series of hypotheses that we believe explain the foundational principles of the design and logical operation of any evolved system for the guidance of intelligent behavior. Our claim is that the most basic problems of life drove the foundational adaptive design of these systems. And whether we're talking about bacteria, bees, birds, bears– or humans–the problem-solving logic of the system's adaptive design is the same.

What do we mean by an adaptive design? Think for a moment about two different types of visual systems–your camera-like eye, and the light-sensitive patches of cells that serve as a kind of eye for the snail. The function–the job description, if you like–of these light-sensitive cells is the same: to convert electromagnetic sensory input into adaptive behavior (which basically means advantageous or fitness-enhancing behavior). Whether it is a snail with a light-sensitive spot, a fly with a compound eye, or a human with a lens and cornea, the underlying architecture serves the same basic purpose and reflects the same basic design feature.

The eye, then, is an example of an adaptation–natural selection's version of how to build a better mousetrap. Different designs are tested for thousands of generations in Mother Nature's school of hard knocks. Only those that survive this relentless winnowing–and pass on via their DNA their recipes for success to their offspring–make the cut. In a sense, our bodies are a collection of natural selection's greatest hits: hearts specialized for pumping blood, livers specialized for removing toxins, hands with opposable thumbs specialized for grasping–and eyes specialized for seeing.

The plastic tissue of the neocortex is also an adaptation. Just as the camera eye you are using to read this paragraph reflects the same purpose and logic as a primitive dimpling of light-sensitive cells, so the plastic nature of the neocortex reflects the same purpose and logic as the primitive memory circuits of bacteria. It is the central component of your behavioral intelligence system–the seat of your mind. Like the bacteria's protein circuits and the bee's swatches of plastic tissue, it communicates with sensory and motor systems to achieve your goals. (What establishes the goals? The part of your intelligence system ultimately responsible for energy management–your life history regulatory system, which we introduce at the end of this chapter.)

But where does psychological experience come in? As we move from the system characteristics shared by all life forms to the more complex intelligences of social animals, psychological phenomena and social dispositions–what we call personalities–begin to appear. Let's go back to the foundational problem of how organisms balance their bioenergetic budgets.

Suppose that you're on vacation in the Western United States—perhaps in Yellowstone National Park in June. This is grizzly bear territory. A subspecies of the brown bear, *Ursus arctos*, the grizzly stands up to seven feet tall, can weigh more than one thousand pounds, and in short bursts can hurtle along at thirty-five miles per hour. (Think of a head-on collision with a compact car that has fur, teeth, and claws.) Although he does eat meat, about 75 percent of his diet is vegetarian—berries, flowers, grasses, nuts, roots. The rest is fish, other animals, honey—and, unless you are careful, the foods you have with you. During the summer, in preparation for six months of hibernation, grizzlies will eat eighty or ninety pounds of food a day if they can get it.

When a grizzly bear is lethargic and lying in his cave, he's likely to be sated—in the psychological state that accompanies being bioenergetically flush. When he's restless and surly, and his radar has locked onto your campsite, he's probably hungry—in the psychological state that accompanies decreasing resources. His energy meter is running dangerously close to empty (and *you* may be the closest 7-Eleven food store). These changes in bioenergetic state exert a powerful influence on behavior, disposition, and psychological experience.

Having an intelligence system that coordinates psychological and behavioral effort makes good adaptive sense. An animal that is motivated and moving about is obviously more likely to find food than a grizzly couch potato: Bears can't call cave service and order up eighty pounds of salmon, moose calves, assorted berries, roots, and nuts, with a side of ants and honey. Think back to the garden. Plants do get a kind of room service—the sun beams in energy, nutrients are supplied in the soil, moisture comes from the atmosphere. Of course, if supplies fail catastrophically, the plant has a major problem. What can it do? The options are limited: wilt, jettison some leaves. But it can't simply pick itself up by the roots and relocate. It dies. End of story. On the face of it, having the ability to move, like the grizzly–to walk away from a bioenergetic disaster and replenish your stores–looks like the more attractive option.

But foraging isn't just a stroll in the woods: There's a kind of energy catch-22 once we introduce behavior to an intelligence system's repertoire. The effort to acquire and conserve energy actually *takes* energy–a lot of energy. Lunch in the wild doesn't come in a package with a bar code. Every opportunity to acquire energy has to be assessed: What are the costs and benefits? Is this a good place to find berries? Is there somewhere farther down the travel lanes of the home range that might yield a better return for the expenditure of the energy needed to get there? What happens if lunch has four legs and fights back or runs? Does the effort of tracking and killing it yield a better net return than digging up tubers? Suppose the bear encounters another animal intent on the same source of food. Again, the energy equation changes and with it, the bear's disposition. It's still hungry, and it's still motivated. But now how effective is the take-no-prisoners disposition it has been displaying? The bear must make a judgment call: Based on its assessment of the other animal's fighting abilities and the value of the resources, does an aggressive encounter make sense?

What is the bear "thinking" at this point? What's going on in his intelligence system? The critical energy equation is automatically being calculated at the level of neurons.

At various sites in his digestive system, there are sensory receptors that recognize glucose. These receptors transmit information about the availability of this readily available form of energy to central energy-management centers in an area of the brain called the ventromedial nucleus of the hypothalamus (VMH). Once the VMH has been informed that on-tap bioenergetic resources are dwindling, it puts the body on alert by generating the psychological experience of hunger. At the same time, a message is sent to another area of the brain, the amygdala. This initiates the associated emotional/dispositional response. Yet another message is sent to the basal ganglia (a large group of nuclei buried deep within the cerebral hemispheres), which begin to select the appropriate motor responses. Finally, a signal is sent to the prefrontal cortex, which begins to plan the foraging excursion.

This is just the list of mechanisms that instigate the energy-acquisition process (and a short list, at that). Once the grizzly has located a food source, he has to secure it and consume it, and these projects require the concerted engagement of several other sets of mechanisms. And if the bear happens upon a competitor for his next meal, a functionally distinctive set of circuits are engaged. One of these signals travels to the sympathetic branch of the autonomic nervous system, generating a battalion of emotional and physiological responses that enable the bear to either fight his competitor or flee the scene. (Notice how elegantly coordinated all this activity is, by the way. It is the product of an integrated system that evolved in accordance with the laws of energy—not of a motley collection of "instincts" and brain modules somehow shunted together to create behavior.)

But what part of this system ultimately determines the

bear's behavior? The neural circuitries we just described are all involved in the process, but there are customized circuits—the individual's unique cost/benefit analysis system—that make the final decision.

All behavioral decision-making systems assess costs and benefits. A behavior has to be worth it.

Think about how you make a simple behavioral decision. (By "you," incidentally, we're not suggesting the existence of a homunculus—a little person inside your head who makes decisions. Rather, we mean the interplay of intricately coordinated hormonal, neuronal, and biochemical systems operating largely beneath conscious awareness.) Let's go back to Yellowstone. You're camping in the wilderness, lying in your tent at night, and you begin to feel hungry. Your stores of bioenergy have dwindled, and your glucose receptors are close to running on empty. You're in grizzly country, remember, so you've secured your food in the back of your SUV, which is a hundred yards away from your tent. Question: Do you leave the warmth of your sleeping bag and the security of your tent, and trek to your vehicle to get a snack? The answer, of course, is that it depends on how hungry you are, and how concerned you are about the odds of running into a bear on the way.

The neuroeconomical networks that support decision-making processes will be described in the chapter that follows, but here's a preliminary explanation of the system logic in this foraging-choice scenario. In humans, like other animals, decreasing glucose levels serve as signals to the intelligence system that readily available energy sources are diminishing. When your intelligence system senses this happening, it begins to calculate the benefit of acquiring food as a function of the degree to which energy is needed. How long has it been since you last ate? How much energy did you expend in setting up the campsite? Are you feeling physically weak or lightheaded? Is this an emergency? It also calculates the costs of acquiring food—the amount of energy that the behavioral effort will require. You're going to have to trek to your vehicle—your system has an estimate of how much effort that will take, because you've already done it once, and that information has been encoded. What are the risks? What's the likelihood of encountering a bear? Your system makes predictions based on past experience, hearsay, and so on. Do bears sleep at night? As your energy reserves continue to diminish, the system's estimates of the benefits of acquiring food increase, while its relative estimates of the costs and risks decrease. When the balance starts to tip in favor of the estimated benefits, a concerted energy-acquisition effort is initiated—and you head for your vehicle in search of food.

If making explicit just a few of the questions that your intelligence system is fielding in this situation sounds oddly mechanical, that's hardly surprising. Most of us—unless we spend time introspecting or talking to psychologists and other species of therapist—are rarely challenged to unravel the processes that lead to our decisions. If someone asks us why we did something, we tend to supply simple reasons. "I was starving, and I figured that the odds of a bear also being hungry at 11 P.M. were pretty remote." But this is just the end result of immensely complex and highly integrated activity by your intelligence system. Not all of it, incidentally, is hidden from view. As you considered the scenario we just examined, you may well have been aware of a succession of representations rafting along on your stream of consciousness—a grizzly bear, images of exactly where your food is stashed, a rehearsal of how quickly and efficiently you could reach it, where you would put your flashlight, whether you would carry some kind of stick in self-defense, whether your cell phone gets reception up here, where the first-aid kit is, and so on. And, of course, your inner voice—the one you can hear as you read these words—would be supplying commentary. In a sense, these are the whisperings of your intelligence system.

## LIFE HISTORY REGULATION: WHAT ARE YOU DOING THE REST OF YOUR LIFE?

There's another, crucial component of your intelligence system that we discuss in detail in Chapter 6: the life history regulatory system (LHRS). In addition to bottom-line maintenance, life has an agenda of higher-order, sequential goals—major construction projects such as development, sexual maturation, and reproduction—that have to be scheduled. The agenda for this is in your genes—but it is flexible. These projects can't be launched without taking your individual life and your environmental circumstances into account. They're energy-expensive projects, and the LHRS is a strategic manager. It makes energetic trade-offs, allocating energy among competing goals.

For example, suppose that you survive your anxious June night in Yellowstone. The grizzly, unknown to you, was sound asleep after an assignation with a female grizzly. (This is probably not a scenario that your agitated intelligence system even contemplated.) And let's suppose that she is now a pregnant bear (May and June are typically mating months for grizzlies). What happens next in her life? A remarkable LHRS-controlled phenomenon called delayed implantation.

Her egg, now fertilized, divides a few times, and the embryo floats unattached in her uterus for about six months; all development stops. At some point in the fall–roughly November–grizzlies head for their dens and hibernate. The embryo then attaches itself to the wall of the mother's womb and continues to grow. After about eight weeks, while the mother is still in hibernation, the cub is born. There is a caveat–and this is why delayed implantation has evolved in this species. If the mother has insufficient fat reserves to survive the winter (and, of course, eventually succor a newborn), the embryo fails to implant and is reabsorbed by her body. For grizzlies, this is the survival of the fattest. In the language we've been using in this chapter, the female's intelligence system (principally, in this case, the LHRS) assesses her bioenergetic reserves and, like a wise but firm bank manager, declines her mortgage on the future. In a sense, it refuses to underwrite her life-stage transition from single female to mother. (There is a kind of parallel in humans: Females who maintain low body fat–like professional athletes–have difficulty conceiving. And there's another side of the coin. Pubescent females with high fat reserves tend to become fertile at a younger age.)

You can see the powerful role that the LHRS plays in the bioenergetic/market dynamics that we have discussed. At what point should an organism reach sexual maturity and begin to reproduce? Should it conserve energy and develop further, or mature quickly, possibly getting into the gene-leveraging business earlier? But what if early reproduction leads to smaller offspring–or maybe to a reduced likelihood of living long enough to nurture them?

There are equivalent questions for humans: What effect does the presence or absence of a father have on the timing of a daughter's menarche (the first menstrual period of a girl in puberty) or on the aggressive behavior of a son? What effect does having a mate have on the timing of onset of menarche? When is it best to be sexually aroused or disinterested?

No generic one-size-fits-all behavioral repertoire is able to optimize energy resources in each life stage. The strategic decisions made by a child's intelligence system are radically different from those made by that of a postmenopausal grandparent. The solutions to the problems of survival, growth, and reproduction vary from life stage to life stage and–we can't stress this enough–from individual to individual. The LHRS orchestrates all of this activity, taking you and your life into account.

## A MIND OF ONE'S OWN

Here's a brief summary so far. Your mind is a product of your evolved behavioral intelligence system, which functions as a customized cost/benefit analysis system. It is modulated throughout the life span by the LHRS, which regulates and schedules the sequential acts of our lives using a genetically encoded script as its guide–but is always responsive, moment by moment, to environmental pressures. And underpinning all of this is a fundamental requirement to meet the bioenergetic bottom line.

Remember that what we have described here is a prototype of a self-adapting system. It is designed to take into account—at every level of its operation—your intricately detailed and absolutely unique life history. It is a system that gives rise to human minds that are unique by design.

# 2

## THE MARKETS OF YOUR MIND

Imagine another garden, in difficult times. The year is 1864; the place, Jonesboro County, Georgia. The plot of land we have in mind lies in the sixty-mile-wide swath of destruction left by Sherman's troops on their infamous March to the Sea. It is the war-ravaged kitchen garden at Tara, the plantation setting of David O. Selznick's film production of Margaret Mitchell's epic novel, *Gone With the Wind.*

Pillaged and burned by Yankee troops, the land lies scorched and barren, save for a few root vegetables. Enter the heroine, Scarlett O'Hara. Weak, hungry, and disillusioned after a series of shattering tragedies—from the destruction of the South to the death of her mother and her father's slipping grip on reality—she walks wearily out to the garden in search of food. Through an opening in the charred fence, she scans the desolate soil for any sign of life. She sees a withered leaf and claws into the clay with her bare hands—only to come up with a pale, dirt-caked radish. She devours it—

then retches it up. In despair, she drops, sobbing, to the ground.

Slowly summoning her strength—music up—Scarlett determinedly pushes herself to her feet and raises a clenched fist toward the heavens. "As God is my witness," she vows, "as God is my witness, they're not going to lick me! I'm going to live through this, and when it's all over, I'll never be hungry again—no, nor any of my folk! If I have to lie, steal, cheat, or kill! As God is my witness, I'll never be hungry again." (Pull back to wide shot of Scarlett silhouetted against the beckoning sky of a new dawn, then fade to black. Cut to scenes of a tough-minded Scarlett single-handedly driving the phoenix-like rise of Tara from the ashes. Gone is the honey-rose Southern belle—a persona that had once served her purposes.)

Scarlett is a consummate player in the human biological marketplace. The critical circumstances of life are fluctuating wildly for all Southerners in the years surrounding the Civil War, and survival hangs in the balance of daily decisions. What becomes of these people when their civilization breaks up? Ashley Wilkes, the object of Scarlett's unrequited love, put it simply: "Those who have brains and courage come through all right. Those who haven't are winnowed out."

But there are a variety of ways to "come through all right." Figuring out how to survive in a biological market depends on many different factors, all changing in real time. And each member of the market is using an individualized mind to make those choices. Some of Scarlett's associates cling to their former social position and "sense of self," adhering to standards of conduct that belong to a world that no longer exists. Other individuals shift alliances abruptly in order to exploit new opportunities. Some, like the film's protagonist, Rhett Butler, use the simple strategy of playing both sides for personal profit, regardless of how the power balances shift over time. And some agile agents, like Scarlett, assiduously track their fluctuating opportunities and form new alliances with each minor change in the market—determined that they'll "never be hungry again."

Although most of us are not coping with the chaotic and dangerous circumstances of war, we all live in a social world in constant flux. We are all individual traders—unique agents of our selves—in the ever-changing markets of life. In a very real sense, we "sell" ourselves as partners, seeking to forge alliances with other agents—potential mates, friends, colleagues, employers. And we "buy" the same commodity from them. We engage in these complex trade interactions, all of the time, all of our lives.

Getting the physical and social goods we need—the resources recognized by our intelligence systems as improving our lives—depends on our ability to negotiate with other agents. It depends on our ability to assess them (as cooperators, for example) and on their ability to assess us. How can we best forge an alliance—or break one? On the surface, this might seem a relatively simple task. In practice, it is complex.

You can be involved in more than one social deal simultaneously—some cooperative, some competitive. A cooperator for one venture may be a competitor for another. Each of these individuals is likely to be involved in a matrix of negotiations just as complicated as yours. And at any moment, your value as a player in this Byzantine social web can change as a function of various factors—some that reflect a change in yourself, your primary cooperators, or competitors (an illness, injury, or pregnancy, for example), and some that result from a change in the greater market (a shift in alliances, a change in the social power structure, and so on).

If all of this sounds abstract, stop for a moment and think about your own life. Which individuals do you value the most as cooperative partners? Whom do you trust? How much do these individuals value and trust you? Which people would you gladly try to help if they asked for it? Which individuals are likely to help you? Can you always count on them? If not, why not? How do you feel about these people? How do you think they feel about you? And how are your relationships with these people affected by the passage of time, ongoing experiences, and changing circumstances?

Suppose, as the result of a downturn in the economy, your company eliminates one of two posts. You are retained; your colleague gets a pink slip. Does your sense of your own value change? How does that affect your relationship with your colleague? Or suppose that your closest friend begins spending time with a newcomer to the community—someone with whom she shares more interests. Does that mean she values you less?

We are intrinsically social animals. Because of our evolutionary heritage, our lives are dependent on our interactions with others; they make sense only in relationship to others. Not surprisingly, the activity of our intelligence systems—our minds—reflects this. If you doubt this, try to imagine a life of utter and unending solitude: Robinson Crusoe with no prospect of ever seeing another footprint in the sand. Try to clear your mind of all reference to other people. Expert meditators are said to be able to do this, but for most of us, the gossip of neurons is simply irresistible. And any attempt at solitary reflection is quickly sabotaged by a parade of images and verbal messages—a mindscape densely populated with representations of the people we have dealings with: family, friends, colleagues, competitors. We call to mind their faces. We "hear" their voices, recall their actions, and revisit our responses to them.

This daily traffic of our dealings with others is inescapable. Were we considerate? Did we cause offense? Was another player taking unfair advantage? Has the relationship been damaged? In our imaginations, we rehearse the likely consequences of decisions. Although, as we said, it takes place largely out of conscious awareness, this is high-priority work for our intelligence systems: Our success or failure in the game of life depends on it.

To say that humans are social animals doesn't mean that humans like to socialize: Rather, it means that we *have* to socialize, if only minimally, in order to survive and thrive. So how do we do it? How do we negotiate with others?

Essentially, our intelligence systems—our minds—construct representations of our self in relationship to other individuals in the social world. They capture and assess the salient information that we gain from our dealings with people, and their dealings with others. And they maintain and update—in real time—a mental register of these transactions. In a sense, our minds are constantly playing a form of social chess—but this is Xtreme chess, played on an ever-changing neural chessboard, with ever-morphing and changing pieces.

For example, what runs through Scarlett's mind as she

contemplates a move to get the money she needs to pay the taxes on Tara? Suppose she seduces and marries her sister Suellen's fiancé, Frank Kennedy? What are the possible consequences? Obviously, Suellen would be crushed—but she can be quickly discounted as an inconsequential market force. More important, Scarlett runs the risk of alienating the honorable Ashley, the longtime object of her love. (Of course, Ashley is married to Melanie, who is ill and may die—which would leave Ashley available to marry Scarlett in the future. What are the odds of this happening, and when?) And then there's the risk of ruining her reputation at large. But how significant would that be if the social merger brings her enough money, and all the power that money can buy?

What are her alternatives? Rhett, the wealthiest member of the market, loves her and should be easily enlisted to her aid. But he's currently in jail, and his funds are temporarily frozen in Liverpool (she has already discovered this). Most of her pre-war suitors are dead. Her best bet is to seduce Frank. From the moment Scarlett determines that Frank has enough money to solve her problems, to the moment she makes her decision to begin seducing him, takes just thirty-five seconds of film time.

An essential component of an intelligence system that is capable of performing this kind of high-speed mental chess is a plastic neural information matrix, the neocortex, where uniquely personal data is retained (and reconfigured) in the form of representations.

What do we mean by a representation? Very roughly, it is the customized pattern of activity of your brain cells that memorializes (represents) the interaction between an object (such as another individual) or an event, and its meaning to your intelligence system—its consequences for your viability at a given moment in your life.

When you see a face across a crowded room, for example, certain neuronal patterns—representations—correspond to your experience of seeing the face. Obviously, this coded pattern is no more the face than a map is a territory or a menu description is food. But it is capable of evoking the image of the face, and all the sensory, emotional, and cognitive associations that accompany it—where you were, what you were doing, and what it meant in terms of your life.

Your representation of the face, and the perceptual experience it creates, is unique to you. This may seem counterintuitive. After all, since the architecture of our intelligence systems and our sensory equipment is pretty much standard-issue (or species-typical), our perceptions of objects in our environment are, for all practical purposes, the same. Well—yes and no.

It's certainly true that your visual system, for example, cannot choose, on-line, how to respond to the light spectrum. The human retina is sensitive to wavelengths within a range of 400 to 700 nanometers (billionths of a meter). We cannot naturally trespass beyond these boundaries—a consequence of the evolved adaptive mesh between the regularities of the world (in this case, the visible part of the electromagnetic spectrum) and a sensory system that reflects those regularities. And so we might assume that we are all seeing, say, a rose in the same way.

But that's not really correct. We certainly have very similar images in our minds of the rose's shape, its color, its scent, and so on. In that sense we share some consensual reality of the rose. But the fact remains that your internalized images of the rose, your customized representations, are the result of a unique information transaction—an engagement between the rose and your individual intelligence system. It is simply not true to say that a rose is a rose is a rose. A rose is also what you make of it, how you perceive it, and what it means to you. A rose given on a wedding anniversary by one person does not have the same meaning as a visually indistinguishable rose left at the site of a fatal traffic accident by another. The representations are unique, unavoidably imbued with each individual's personal histories and emotions. And so your gift of a particular rose becomes part of your personal representational history, which is being dynamically constructed and updated in your neocortex in each moment.

As well as being unique to us, our representations of objects and people are constantly being modified by our experiences and by changes in the social world. Frank Kennedy's face evoked a very different representational pattern in Scarlett's intelligence system when she saw him at the barbeque at Twelve Oaks before the war than when she saw him in Atlanta during Reconstruction. His face had changed little in the years that had passed. What *had* changed was the biological marketplace and their relative positions in it—and, most important for our purposes, the corresponding representations in Scarlett's mind. Her intelligence system had updated itself with every biological market change that had impacted on her over the years. It held an up-to-date representation of herself and her position in the market at that moment—a poverty-stricken, socially displaced former

Southern belle, now desperate for a few hundred dollars. It also held a newly created representation of Frank as someone with a thriving business, a foothold in the new order of the market, and—consequently—the ability to solve her problems.

We aren't born with a neural representation of the world we live in. It takes time for our brains to develop the foundational neural tissue that will eventually become the scaffolding of the self. In evolutionary terms, this makes sense; our immature cortex hasn't been taxed with solving a survival problem. In the womb—and to some extent as an infant—you have no need of a mind that can navigate a biological marketplace teeming with potential collaborators and competitors. In the primal economy, you are dependent on transactions with only one person: your mother.

You might imagine that the womb would be the ultimate sanctuary, off-limits to the seemingly callous calculus of costs and benefits that characterizes life in a marketplace. But you would be wrong. Although you were not yet a member of the social biological marketplace, you were engaged in serious physiological "transactions" with your mother well before you were born. Before you opened your eyes for your first unfocused view of the world, before you even uttered your first cry, you were already a seasoned veteran of the markets of life.

In the womb, you were engaged in a struggle for survival. In this supposedly tranquil and privileged place, as you floated tethered by an umbilical cord, you were striking biochemical bargains, negotiating for more nutrients from your mother. Why? Because life requires energy, and when the same two individuals are tapping a limited supply, competition results. No two individuals have precisely the same interests, the same goals, the same strategies for survival—not even a mother and her fetus.

Here's the explanation. Let's go back to the basics of natural selection. Natural selection promotes, or favors, genes that increase the survival and reproductive ability of the individuals who carry them. Over the immensely long haul of evolutionary history, this has been a recipe for success. Individual life forms that were endowed with characteristics that enabled them to survive environmental challenges and successfully reproduce passed on those traits to their offspring. The universal aspects of human nature are a broad composite of shared physical, psychological, and behavioral characteristics, and they exist precisely because these characteristics were effective in promoting the survival and reproductive success of our ancestors. For social creatures, like humans, whose offspring take some time to mature, inherited characteristics that promoted the well-being of offspring became part of our genetic heritage.

This does not mean, incidentally, that human beings are compelled by ruthless "selfish genes" to go forth mindlessly and multiply. In fact, natural selection has evolved a far more subtle approach—a kind of genetic sting operation that exploits our response to pleasure. Our intelligence systems are genetically predisposed to respond positively to sex, to social engagement, to parenting, for example. By and large, when we engage in these activities, we are rewarded by feelings of pleasure. And so we repeat them—which leads to the passing on of genes.

Genes are a family affair. You get fifty percent of your

genes from your father and slightly more than fifty percent from your mother (an interesting technicality: The ovum carries some genetic material outside the genome; the sperm does not). Your siblings also get fifty percent of each parent's genes. As you move out to the further reaches of the family, the degree of relatedness attenuates and the bloodline thins. And evolutionary processes have favored genes that give rise to physical, psychological, and behavioral characteristics that tend to promote the well-being of your genetic relatives, as a function of your degree of relatedness to them, rather than to strangers or even friends. (Although these fifty percent inheritances sound very large, remember that the variation in *all* human genomes is vanishingly small: We share 99.9 percent of the sequences of our genetic code.)

As we talk about these mathematically rigid genetic co-efficients of relationship and their effects on our psychology and behavior, it's important to enter a caveat. For humans, certainly, these are not immutable laws: They are malleable predispositions. The awkward fact remains that individuals can elect to bequeath their entire estates to establishments supporting the care of cats and dogs, rather than relatives. Individuals can elect to favor friends rather than family members, perhaps investing time and resources in the offspring of colleagues rather than cousins. Individuals can become foster parents or adoptive parents—essentially underwriting the genetic legacies of often-unknown individuals (frequently from different ethnic backgrounds). And individuals can become painfully estranged from their own kin. (It is a matter of record, for example, that Abraham Maslow—ironically, one of the founders of humanistic psychology, whose theories we explore in Chapter 8–had a disastrous relationship with his mother and could not bring himself to attend her funeral.)

With the exception of identical twins, there is no greater degree of genetic relatedness than that between a mother and her child. That is why mothers usually give so "unselfishly" to their offspring: It is in the interest of a mother's genes that her offspring survive. And that is why mothers are, in a sense, programmed to provide for all of their offspring's bioenergetic needs during the early stages of development. This initial parental investment is the best deal a human will get in the biological market for the rest of his or her life.

The initial bioenergetic deal a human has with his or her mother is purely physiological: Mothers transfer their own energetic resources to the fetus through the placenta. This is part of the ancient and largely well-regulated process of reproduction and pregnancy. Smoothly and automatically, hormonal battalions are mobilized, supply lines are established, maternal feelings are promoted and experienced, and the scene is set for birth.

Although this transformation appears to be a selfless gift, evolutionary biologists have long argued that this arrangement evolved because it serves the agenda of the mother's genes. She is provisioning her own genes (packaged for posterity in the fetus) with the goods needed to survive into the next generation. There is one obvious caveat: She needs to reserve enough of her resources to sustain her own existence and allow for the possibility of future offspring. If she were the sole player in this scenario, her system could make unilateral decisions, allocating resources as needed to achieve her goals. But she isn't.

Evolution has equipped the apparently helpless fetus with a few tricks for siphoning off more of the goods than its mother's system is genetically programmed to supply. Even in this first cooperative relationship in the biological market, there is a degree of competition—a conflict between the mother and the fetus that arises from the differences in their genetic makeup and, therefore, genetic interests: maternal-fetal conflict.

In maternal-fetal conflict, the budding intelligence system of the fetus negotiates aggressively for additional resources by manipulating the mother's physiology to its own benefit. (It's important to reiterate that there are no conscious motives involved here—just physiological consequences of a genetic conflict of interests.) And this is where the embryonic life history regulatory component of the fetal system comes into play. Cells derived from the life history regulatory system (called trophoblast cells) invade the wall lining of the mother's uterus. There, they remodel the uterine blood vessels so that they are unable to constrict (preventing the mother's system from regulating the supply of nutrient-rich blood to the fetus). As a result, the fetus is able to hijack extra resources.

Because the vessels can't constrict, the fetus can now tap into the mother's blood supply. The remodeling also raises the mother's blood pressure, which forces an increase in the volume of blood (and the bioenergetic goods it carries) coming through the placenta. Finally, the fetus is able to inject its own hormones through the remodeled vessels into the mother's blood supply. One of the hormones it injects (human placental lactogen, hPL) inhibits the effects of insulin, a hormone produced by the mother that enables blood sugar to be taken up by the cells and used as fuel. Why does it do this? Because the fetal system is hungry for more glucose and "wants" to keep the mother's blood-sugar levels high. The mother's system, on the other hand, wants to use the blood sugar. But it can't–because its insulin has been rendered ineffective. And so maternal blood-sugar levels can get dangerously high (characteristic of diabetes, which is a risk factor in pregnancy). And so it goes, with dueling hormones acting on behalf of the competing players. Even in this most cooperative of places, where the future survival of both parties is inextricably intertwined, biological market forces are at work.

After birth, what then? In most societies, in most households, your mother (or it could be another primary caretaker) continues to mediate your life needs as you begin to construct your initial self-representations. You are now embarking on the long childhood–the extended period of parental care that exceeds that of any other animal. Your mother has already invested a huge amount of energy in bringing you into the world. And now she faces a feeding/weaning period that could last for two to four years. Every single day, she has to supply you, through lactation, with a substantial amount of energy–in addition to keeping herself alive.

This is a major undertaking–but year after year, generation after generation, the evolutionary bottom line is that natural selection has supplied both you and your mother with physiological, psychological, and behavioral mechanisms that prompted a ready transfer of her resources to you–along with a repertoire of maternal mechanisms that motivated her to protect you and teach you how to survive.

For example, you were born with a sucking reflex–involuntary sucking movements of the muscles around your mouth in response to stimulation (your mother's nipple). Sucking sends signals through the sensory nerves of the nipple to a mother's brain. The signals then pass upward through her brain stem to neurons in her hypothalamus, leading to the release of the LHRS hormone oxytocin. Oxytocin is carried by the blood to the breasts, where it causes contractions of special cells (called myoepithelial cells) that form a latticework around the alveoli of the mammary glands. The result? Less than a minute after an infant begins suckling, mother's milk begins to flow–an innate behavioral mechanism in the infant meshing with an innate physiological mechanism in the mother to ensure that the infant's bioenergetic needs are met.

Once the child is born–at least in theory–a mother could refuse to serve any longer as a volunteer energy-delivery system. Usually she doesn't. Why not? The commonsense answer is that she loves her child. But the ultimate reason is this: The child is her message in a bottle, her (genetic) hopes and dreams tossed into the ocean of time. And the proximate reason, her feelings of maternal love, is simply a part of Mother Nature's sting operation: Oxytocin, released in the period around the time of the child's birth, induces the feelings that motivate nurturing behavior.

Oxytocin is a powerful potion. Sometimes–inaccurately–called the love hormone, it is thought to be responsible for orchestrating maternal and affiliative reponses. Here's how effective it is. In wild mice raised in the laboratory, pregnant females normally cannibalize any babies (pups) they encounter before they give birth to their own pups. But the closer they get to the moment of birth (in other words, the more oxytocin is bathing their systems), the less likely the behavior. In fact, if they are injected with oxytocin while they are cannibalizing pups, they abruptly stop the attack and shift into maternal mode.

In human studies, women with high levels of oxytocin report feelings of greater calm and evidence lower blood pressure and lower levels of anxiety, together with greater attachment to their children. They breast-feed longer and enter into what might be thought of as a neurochemical compact with their babies. Some scientists have even suggested that oxytocin and another hormone, vasopressin, might be thought of as bonding hormones that lead to the formation of alliances, which is the very foundation of social behavior.

Even so, there are limits to a mother's selfless provisioning of one of her children. Do the math. What does it cost to be the primary supplier of resources to an only child? Remember, we're talking about how these traits evolved in ancestral times. Forget kindergarten, home help, supermarkets, and car pools. Think about the African savannah. Think about foraging daily with an infant attached to your breast. Think about the fact that continuous nursing is a kind of prophylactic: It is a physiological barrier to conception (and a natural method of spacing births to three or four years apart). You may want to have another child (if resources allow it). But your firstborn has a different agenda. Given the chance, mammalian offspring—including human infants—will continue to suckle long after it is necessary for survival.

Your firstborn carries half your genes—but so will a second child. From your perspective, it makes sense to wean the firstborn as soon as possible and perhaps double your genetic investment with another child. But now there's a problem. Although the first offspring would derive some genetic benefit from the birth of a sibling (it would carry fifty percent of his/her genes), it comes at too great a cost—loss of the ability to get nature's free lunch: mother's milk. The stage is now set for parent-offspring conflict. Parent-offspring conflict over weaning can last for weeks, or even months, with the firstborn infant throwing tantrums that invoke feelings of guilt in the mother. In some species, like baboons, weaning conflict often involves physical violence, with the infants lashing out at their mother.

Once a mother has successfully weaned a first infant, her physiology changes and she is much more likely to conceive another child. This is the prelude to yet another genetic conflict: sibling rivalry. Let's go back to the gene calculator. Parents share fifty percent of their genes with all their biological offspring. So, in theory, parents should place approximately equal value on all their children.

A child's assessment of his or her sibling's value is very different. On average, full siblings are fifty percent genetically identical to each other—which is the foundation for a strong bond. But there remains a fifty percent difference. To the extent that genetic relatedness influences psychological and behavioral realities, a child should consider itself to be roughly twice as valuable as any of its siblings.

As a consequence, siblings tend to compete with each

other for the emotional and financial investment of the parents. Despite the frequency and intensity of any at-home rivalry between them, siblings should nevertheless tend to cooperate with each other if it is the only way for any of them to get the parental goods. And should an outside party threaten the safety or well-being of an individual's brother or sister, intense loyalty and defensive behavior are likely to emerge. Ultimately, just as sibling conflict is fueled by genetic disparity, sibling cooperation is fueled by genetic similarity.

All of these early familial social interactions, cooperative and competitive, were guided by your intelligence system. As you steered a course through the ongoing encounters, your neocortex was developing increasingly complex customized representations about your social and physical world. Bootstrapping off your sucking reflex, you began to develop network associations that broadened your behavioral repertoire, enabling you to take milk from a bottle, then a cup. At the same time, you began to develop networks of information that represented the important people in your life and your relationship to them: representations about what and whom each person valued, what they knew of their world, "mental theories" of the contents of their minds.

Just as your brain was beginning to form an individualized functional map of the physical world, a map that enabled you to predict where you could find what you needed and what to avoid, it was also beginning to form a functional map of the social world. It was creating neural/mental representations of the nature of the relationships between the important people in your life, representations that enabled you to predict how these individuals were likely to behave in various situations, and how their behavior might affect your life. It was recording, in an on-line fashion, the ongoing changes in these relationships. And of course, this multi-dimensional virtual map of the social world—the representational biological market in your mind—would be of little use unless it contained a virtual representation of you: your "self."

But how does your intelligence system actually create these representations of your physical and social world and, ultimately, representations of your "self"? As we said earlier, the key lies in the neocortical memory component of your intelligence system and its remarkable functionally plastic properties.

Let's go back to Chapter 1 for a moment to the rudimentary memory component of the bacteria. The intelligence system of the bacteria has receptors for glucose (the bioenergetic "goods"); it has a four-second memory so it knows whether glucose concentration has just increased or decreased; and it has a behavioral effector system that will propel it in the direction of the highest glucose concentration in recent history. The bacteria's memory component allows for an encoded association between resources in the environment and a best guess at what the most adaptive path of behavior should be.

Here's a slightly more complex situation. Recall our garden from the last chapter. The primary biological market business is pollination. Different classes of traders—flowers, bees, hummingbirds, and so on—exchange commodities. In this case, the principal commodities are a sexual transport service (carriage of the flower's pollen, which contains its genetic information) and bioenergetic resources (provisioning of food—mostly glucose-containing nectar). Ultimately, the function of a flower is to seduce one of the various pollen-delivery systems into paying a visit, bribing it with nectar. But how does the bee, say, get the most out of its foraging trips? How does it "know" which flowers are likely to have the highest value in terms of nectar?

Basically, bees learn to associate the location, shape, symmetry, color, and scent of flowers with the experience of getting the bioenergetic goods—the nectar. They store this information in the form of representations, and use it on their next trip. Researchers have shown, for example, that bees seem to prefer large, symmetrical flowers. Why? Because large, symmetrical flowers adorn plants that reliably produce larger quantities of nectar. And just as the smell of fresh baked bread can remind us of a time and a place where we had a meal, bees are also able to associate a particular scent with a valuable food source, and the color of the flower.

Suppose a bee makes a landing on a blue flower and starts "sipping" nectar. So now the bee associates "blue" with food. This is a new, learned association. "Blue" did not necessarily signal the presence of food to the bee's ancestors, and the bee wasn't born with an instinctive association between "blue" and food. But the next time the bee is hungry, it will tend to fly toward blue flowers. It paid off on the last flight: Why not again? Now suppose that environmental conditions change. Maybe it's a different season; maybe some climatic factor affects the concentration of nectar in different species of flowers. Whatever the reason, blue flowers begin to produce less nectar and yellow flowers begin to produce more. What happens? The answer is that there will soon be a weakening of the bee's association between "blue" and food, and the development and strengthening of a new association between "yellow" and the food.

What's more, bees quickly figure out the intricacies of shifts in the foraging landscape. Suppose you create a patch of randomly mixed yellow and blue artificial flowers. You doctor the patch so that all of the blue flowers are laced with a couple of drops of sugar (the glucose in nectar). The yellow flowers are treated differently: one-third are laced with six drops of nectar; the remaining two-thirds are left dry. Overall, although differently distributed, the aggregate amount of sugar in the blue and yellow flowers is equal. Now you allow the bees to forage, sampling the blue and yellow flowers and constructing a strategy.

The result? Bees go to the "safe," constant-yield, blue flowers 85 percent of the time. It's as if they do a mental computation that goes like this: Suppose there are 120 blue flowers and 120 yellow flowers. One hundred twenty of the blue flowers have two units of sugar, for a total of 240 units. Forty of the yellow flowers have six units of sugar; the remaining 80 are empty (also for a total of 240 units). If a bee samples all of the blue and yellow flowers, the result is the same. But that's not how life goes in the world of a bee. It has no interest in long-term statistical averaging. It has to make a living and cannot afford to cruise and sample all the flowers, burning energy as it goes, without the prospect of refueling. It needs a reliable glucose fix, rejects the yellow flowers as a bee version of a Las Vegas casino, and concentrates on the blue flowers for its meal ticket. If you reverse the dosages so that the yellow flowers offer a reliable amount of nectar and the blue flowers are a crapshoot, guess what happens. The bees change their strategy and forage overwhelmingly in the yellow flowers.

Bees, like all animals, are energy loss–averse. When their strategy is not working, they revise it long before they are in difficulty. In evolutionary terms, this makes sense and reflects a fundamental asymmetry in the way most organisms evaluate their prospects. At any given moment, avoiding loss (anthropomorphizing, you could say adopting a realistic pessimism) is more critical than increasing a surplus. For a surplus–no matter how large–means that you are safely above a baseline. A loss, however, means you are courting disaster. No organism can afford to persist in behaviors that result in a bioenergetic bank balance in the red. In nature, debt is death. Which is precisely why all mammalian intelligence systems–including ours–are constantly monitoring the energetic costs of behavior and weighing them against the benefits (the goods).

How does the bee track these changes in the environment and successfully navigate toward the goods? Very much in the same way you track changes in your environment and navigate your way through your ever-changing world. The bee has evolved an integrated neuronal system that guides foraging behavior. One part of this system is instinctual, reflecting the regularities of the foraging problem. But there is a new element–a crucial addition to the bee's behavioral intelligence system that enables it to go beyond the "if A, then B" survival logic of *E. coli*. It has a rudimentary cortex–a swatch of functionally plastic neural tissue–that can encode critical information about previously novel features of the environment and how they relate to the bottom-line activity of getting the goods. It enables the bee to learn from past experiences (the blue flowers were full of nectar on the last trip), make predictions about future rewards in variable environments (but they seem to have dried up; maybe I'd better try the yellow), and execute fitness-enhancing behavior on-line (Ah! Much better. I'll take my pollen-delivery service to the yellows for a while . . .). The bee's intelligence system enables it to learn the new market environment quickly and home in on nectar—much as Scarlett's intelligence system enabled her to forage through Atlanta and home in on the person who had the goods she needed.

It may seem a leap from a bee on aerial maneuvers to Scarlett's (or your) navigating a changing biological marketplace—but it's a crucial point, and one that warrants a careful explanation. If you understand the mechanism that the bee uses to learn each new market environment and form associations between various floral features and "the goods," you can understand the basic logic of your mind.

Look at the key components of the bee's guidance program. First, it needs a core instinctual mechanism. An evolved connection between a sensory stimulus in the environment (in this case, the sugary sweetness of the nectar) and a specific behavioral response (in this case, extension of the proboscis—the nectar-sucking tube). For this instinctual mechanism to have evolved, proboscis extension in the presence of the nectar must have reliably led to the bioenergetic goods for ancestral bees. In the early stages of development, your intelligence system relies on an analogous instinct—simply substitute the sucking reflex for proboscis extension, and mother's milk for nectar, and you'll understand what we mean. As you begin to navigate your path in the markets of life, these instincts are the first step in the creation of your customized representations.

In addition to this instinct, the bee needs sensory systems, such as visual and olfactory systems, that can detect and record features of the environment that were previously arbitrary (such as the color or odor of a particular flower) but have now become associated with the availability of food. (Just as your visual system allowed you to form an association between a bottle and milk.)

Third, it needs a plastic neural substrate—like your neocortex—in which these new associative network connections ("adaptive representations") can be formed. Finally, it needs a neural mechanism to build and modify these representations in light of experience. (Remember, these representations automatically self-adapt with the outcome of each relevant experience, increasing or decreasing their strength as a function of the value of the acquired "goods.")

In the bee, the plastic neural substrate is found in structures called the mushroom bodies and protocerebral lobes of its brain—the bee's equivalent of your neocortex. The neural mechanism that forms new representations is a special neuron (called "VUMmx1") that projects from the life history regulatory system (the part of the system that "knows" whether or not the bee just got food) to the plastic part of the brain. When the goods have been registered, this neuron ensures that the sensory information—which just came into the brain—will be associated with the appropriate foraging and eating behavior. (We have an analogous mechanism for creating and modifying these representations, which we describe in Chapter 5.)

Our intelligence system works in much the same way. Our neocortex allows us to form customized representations of our unique experiences. Of course, compared to the bee's plastic brain tissue, our neocortex is huge and is embedded in a significantly more sophisticated intelligence system than the bee's. (Human cortex is about three to five millimeters thick and covers roughly 2,500 square centimeters. It contains about 100,000 neurons per cubic millimeter.) As a consequence, it has much greater representational potential and raw computational power. And yet the underlying logic is identical: You use the same type of adaptive representational networks to navigate your biological marketplace as the bee's brain uses to forage in a field of flowers.

Why do humans have so much of this amazing functionally plastic brain tissue? Ultimately, because we are mammals. The evolutionary appearance of the neocortex in mammals is most likely the consequence of the unprecedented amount of uncertainty in the social biological marketplace—negotiating for mates, and cooperating and competing for food and shelter. The more functionally plastic memory material one of these early mammals had, the more likely it was to survive the vagaries of a fluctuating social environment. And so, over evolutionary time, the outer layers of the mammalian brain increased in size.

Researchers have also shown that, in primates, there is a significant relationship between social-group size and the ratio of neocortex volume compared to the rest of the brain. The bigger the social group, the greater the proportion of neocortex. This is likely to have been the result of an escalating co-evolutionary process. The complexity of the social environment resulted in natural selection for more cortex, which ultimately increased the complexity of the social environment, which perpetuated the selection pressure for more cortex. And so on.

Let's go back to the days when you were beginning to form representational maps of your marketplace. Think back to some of your earliest memories. Chances are you can recall various parental lessons–the social feedback you got about which behaviors in the physical and social world were "good"–likely to increase your viability over the course of your life, and those that were "bad"–likely to decrease your viability over time. Your parents probably praised you when you behaved in a way they found appropriate, and admonished you when you did something they found inappropriate. In "correcting" your behavioral course, your parents were providing feedback to your cortex (via sensory and perceptual systems) that began the construction of your adaptive representational networks–feedback that the world would otherwise provide (although probably much more harshly).

The small rewards they provided for some behaviors built up neural connections that increased the likelihood of your repeating those behaviors in similar circumstances in the future. And small punishments had the opposite effect. This "negative" process is handled by additional mechanisms (primarily in the amygdala and other subcortical structures) that associate these behaviors with painful or otherwise aversive sensations.

These early representational networks–a kind of behavioral training wheels–probably gave you a good start in life. When your intelligence system guided you into the kitchen (because the physical state associated with hunger initiated activity in cortical networks that retained the information that kitchens are where food is likely to be found), it also stopped you short of touching hot objects on the stove. Why? Because the neural signal that was sent as a result of simply seeing the stove activated adaptive representational networks that retained a record of the times when you had reached out to touch a kettle and been startled by a loud "No! Don't touch!", a small slap on the hand, or, in the worst case, burnt fingers.

As you got older, life became increasingly more complex, and individuals other than your parents began to provide feedback to your intelligence system. Your siblings or cousins might have acted as competitors in some of these "training" experiences, providing feedback designed to benefit their viability, rather than yours. As your life progressed and you moved into the world at large, there was less and less consistency in the feedback your behavior received, and more behavioral options available to you. The increased complexity of your new world was reflected in the increased complexity of your representational networks and the decision-making processes that they support.

Think about the role these representational networks play in a simple decision-making scenario. Suppose you're hosting a dinner party, and you're trying to decide where to seat the guests. You consider the first name on your list, and your representational record of that individual is activated. Some of these representations reflect the person's interests, some reflect political ideology, and so on. As you move down the list to the next individual, similar records are activated. Instantly—courtesy of the associative nature of these networks—you recall a party where these two people got into a heated debate. As you continue down the list, your mind begins to associatively create a "model" of each alternative seating arrangement, constructing temporary representations of two individuals interacting. On one dimension, two guests match: Both are single and looking for a partner. You have stored representations of positive outcomes resulting from such an arrangement, so that element of your representational model gets a "plus" (which you experience as a positive feeling). But on another dimension—say, religious affiliation and tolerance—the two guests don't match up well at all. You have many representations of similar interactions that led to social disaster. So that element of your representational model gets a big "minus" (which you experience as a negative feeling, and shudder at the thought). Your intelligence system continues to construct models of potential seating arrangements. Each one has a network value. Your system does a quick real-time calculation, adding up positive values and subtracting negative ones, to come up with a final total for each arrangement, which then gets either a mental thumbs-up or thumbs-down. Your subjective experience of all this activity is that you decided on the option that gave you pleasure.

Not all of life's decisions are so straightforward. Some affect many different aspects of our lives, and, as a consequence, our intelligence system has to perform cost/benefit analyses within and between various different problem-solving domains. For example, suppose you are weighing a decision to take a job in another city. Perhaps the job pays more than the one you currently have, but the cost of living in the other city is also higher (will there be a net gain in discretionary income?). Perhaps your child is in middle school and is well adjusted to his current school environment; a move could lead to social setbacks similar to those that have occurred when major changes in his life occurred in the past. Maybe your spouse won't be able to pursue her chosen profession in the new city, but, on the positive side, she'd be just a short train ride from where her parents live. Although major life decisions can involve multiple mental equations across various distinctive domains, your intelligence system comes to a conclusion by performing the same type of cost/benefit analyses, creating mental models of staying where you are or moving to the new city, assessing the likely pluses and minuses associated with one model versus the other.

Our behavioral intelligence systems encode representations of our experiences and use them to create models of the paths we might take into one or another possible future. Although the system that constructs and modifies these representations is common to all of us, the representations themselves—and the virtual past, present, and future worlds they create in the minds of individuals—are unique. Revisit Tara for a moment, and recall the distinctive possible futures the characters modeled when they heard that General Lee had surrendered. Gerald O'Hara's last bit of hope for the South—and his pride—vanished with the news, and his face reflected a future of shame and poverty. Melanie's mind and face lit up imagining Ashley's homecoming. And

Scarlett—always the opportunist—positively beamed at the prospect of new economic opportunities: "We'll plant more cotton. Cotton ought to go sky-high next year!" Although they lived in the same house and in the same world, their life experiences had been theirs alone, the commodities they valued were distinctive, and the markets represented in their minds were unique.

# 3

## THE SCAFFOLDING OF THE SELF

Laurence Olivier was probably the most respected English-speaking actor of the twentieth century. He acted onstage for nearly sixty years, made almost eighty films and television programs, and directed dozens of stage productions. You might imagine, after this extraordinary career of exploring and portraying different facets of human nature—and assuming close to two hundred identities—that he would have insights to share about the presentation of the self in everyday life. And yet, according to biographer Anthony Holden, Olivier was "a born actor who has spent his long life auditioning to be himself, the one role he could never quite pin down." Olivier was seen by many as "an empty shell, even a hollow man, a vacuum waiting to be filled by circumstance." One of his great contemporaries, Sir John Gielgud, observed that Olivier "is a great impersonator. I am always myself."

Before taking a writer to meet Olivier, theater critic Kenneth Tynan said: "Now, what you've got to realize about

Olivier is that he's like a blank page, and he'll be whatever you want him to be. He'll wait for you to give him a cue, and then he'll try to be that sort of person." And Dame Joan Plowright (Olivier's third wife), when asked how she knew when her husband was acting and when he was not, said, "Oh, Larry's acting all the time."

Of course, this is what actors are supposed to do: They take on different personas, wear different masks. (The word *persona* comes from the Latin *per sonare*–literally "to sound through," or speak through, a mask representing the character an actor was playing.) Olivier was a master of disguising his personality (same word root) by adopting a mask–usually a bit of stagecraft, like a false nose or a strange accent–to convince audiences that they were witnessing a different individual, a novel and distinct self. In that sense, Gielgud's comment "I am always myself" was entirely accurate. Whether he was playing Hamlet or Dudley Moore's butler, Hobson, in the movie *Arthur,* Gielgud was always Gielgud playing Gielgud. (Most of us fall somewhere between these two extremes, with varying abilities to adapt our personalities to different situations.) As with Gielgud, so too with many major movie stars: they play themselves–whatever the plot. And when they play "against character"–against the image that we have formed of them–we are often disappointed, even irritated.

There's a reason for this. As we navigate social space, we are more at ease with people who project a consistent and predictable personality. We know what we're buying. In the U.S. presidential election of 2000–despite his talk of being his own man and of having an irreducible core–Vice

President Al Gore suffered during the debates by projecting different personas and raising questions about who he "really was."

So it's not surprising that Olivier was felt to be a kind of social chameleon—onstage and off—morphing apparently effortlessly from one self to the next. Plainly, the man was doing something right. For his services to the theater, he was awarded a life peerage and the rare honor of the Order of Merit. He was the first director of the National Theatre Company and had a building named for him. And yet it's clear that those who knew him well felt that there was something fundamentally unsettling in his ability to be all things to all people. Their unspoken question was: Where is the *authentic* Olivier—the unchanging core self?

Enter the word *self* into an Internet search engine and you will be overwhelmed with literally millions of references. Sociologists have a thriving literature about the self. So do psychologists. Anthropologists have explored the concept of self—or "person"—in different cultures. Religions offer a different perspective. One example: A fundamental Buddhist idea is *anatta,* or no-self—the notion that our experience of being separate individuals is illusory. And as the vexing problem of consciousness enters the scientific mainstream—largely as a result of new brain imaging technologies—neuroscientists, philosophers, and even physicists have begun to advance theories about the function of our sense of self.

Our approach has been to return to first principles. There are two key questions to be answered: What is the utility of having a self? (What does this powerful and pervasive sense of your uniqueness and individuality buy you in energetic terms?) And—at the foundational level of neurons and their networks—how is a self-representation constructed in the architecture of the mind?

First—obvious—point: Every organism interacts in some way with its environment—from varying concentrations of nutrients to the behavior of other players in a biological marketplace. There are no exceptions to this. Bacteria, bees, and humans have mechanisms for registering changes in their fortunes and altering their behavior. (I've hit the doldrums in the sea of glucose; I need to go into tumble mode. Or: The blue flowers have dried up; I need to try the yellow. Or: I've been passed over for promotion; I need to reconsider my position with this company.) There's a common thread to these problems—problems that must be solved by all life forms: Am I meeting the bottom line?

The story for *E. coli*, remember, is quite simple. It has about four seconds to figure out whether life is getting better or worse, and then it charts a course through the next, brief block of time. Bacteria do not ponder the successes and failures of their adventures in your digestive system. There is no bacterial angst about whether they should have run instead of tumbled, no self-analysis, no ability to recall their experiences and utilize hindsight to plan for the future. But they are assessing where they are in relation to the goods. Bees, with the advantage of plastic tissue and a million neurons, have a greatly expanded memory: Their universe is larger and richer than *E. coli*'s. They have the ability to return to the hive, share information, and lead new expeditions to foraging territories that they have already mapped and explored. This is an astonishing feat, given their limited computational abilities. (Your brain contains a million times more neurons than a bee's.) There is nothing to suggest that bees reflect on their missions. Even so, in a very rudimentary fashion, these intelligence systems set the stage for the prodigious capacity for self-reflection that is the hallmark of our lives.

For us, this kind of reflection is second nature. How is it done? To think about yourself in relation to your world, you need the machinery to construct a self-representation—a neural signature that uniquely defines you as the central player in the marketplace. You also have to be able to construct representations of the other salient features in your environment. Then you'd need to create a relational map in your mind between your self and those salient features—the things in the world that have an effect on you. And all of this has to be dynamic, responsive to environmental fluctuations (like the changing concentrations of nectar in the blue or yellow flowers) and to changes in your internal state.

Let's start with your internal state. Nestled under the relatively recently evolved hemispheres of your neocortex are a number of evolutionarily older structures that include the nuclei (clusters of cells) of the hypothalamus, and various brain stem nuclei. The hypothalamic nuclei are core components of your life history regulatory system. They sit at the center of a web of connections, linked to the ancient brain stem, the pituitary gland (the conductor of the endocrine orchestra), and the neocortex—where your uniquely personal history of experiences is stored. At any moment in time, the state of activation of the hypothalamic nuclei is a neural snapshot—a representation—of your body's internal state, your internal milieu. In a sense, it's an image of what you need in your life at that precise moment.

The main function of the hypothalamus is maintaining homeostasis. As we noted in Chapter 1, homeostasis is the process of regulating the status of the body—measured by, for example, blood pressure, body temperature, fluid and electrolyte balance, pH, hormone concentrations, or blood glucose levels. To do its job, the hypothalamus receives constant inputs about the state of the body from its various outposts, and is then in a position to initiate changes if any of the indicators is off-kilter. By sending neural signals to the autonomic nervous system, the hypothalamus can control heart rate, digestion, perspiration, and so on. By sending endocrine (chemical) signals via the bloodstream to the pituitary, the hypothalamus can effectively control every endocrine gland in the body and alter blood pressure, body temperature, metabolic rate, and adrenaline levels. Because it is sitting in such a pivotal position at the center of its own subcortical empire, the hypothalamus is ideally placed to help the cortex generate the most fundamental kind of self. (Of course, the generation of this fundamental self representation requires a constructional/activational system that intermediates between the two—like the bee's VUMmx1 neuron; again, this system will be described in Chapter 5.)

At any point in time, the activational state of the hypothalamic nuclei defines the behavioral problem that your intelligence system needs to solve. (Problems to be solved in the social world are further directed by the activation of other subcortical structures that guide us emotionally.) Suppose, for example, that your internal state is registering the fact that your body temperature is low. You feel cold. You'll re-member from Chapter 1 that your autonomic system reflex-ively shifts into gear: You shiver (and there are the visible signs of goosebumps). This is a physical representation–a neural snapshot–of a "cold you." Now, suppose that as an in-fant, one of the earliest remedies for your having been cold was your mother pulling a blanket over you. Since then, of course, you've learned the trick so well yourself that you can do it with your eyes shut. (In fact, that's exactly what you do when you're asleep and not even conscious of the behavior.) So in this moment of coldness, you also have a snapshot rep-resentation of something in your sensory environment–in this case a blanket. You also have a third snapshot: a repre-sentation of a behavior (in this case, pulling a blanket over you). And finally, you have a fourth snapshot: the repre-sentation of a behavioral outcome–in this case, you are no longer cold, and your internal milieu (at least in the body-temperature dimension) has returned to acceptable limits.

To continue the photography metaphor, imagine that the four snapshots–the four representations–are spliced to-gether and put into motion. (The neural networks of your in-telligence system are, after all, dynamic, not static.) You will have a brief but meaningful scene from your life.

This is a small adaptive representational network (ARN)–the fundamental unit of intelligence–that associa-tively links your internal state, the sensory features in your environment, your behavioral response, and the adaptive value of the behavioral outcome reflected in physiological and affective changes in your state. Run the movie again. You are asleep, cold, shivering. You reach out for the blan-ket and tug it over you. Moments later you are warmer and feel content. All achieved by the magic of adaptive representational networks.

As your experiences multiply, you build up a dynamic virtual-reality archive of scenes from your life and the adaptive solutions (or failures) to problems you have encountered in the marketplace. The next time you feel cold, for example, your intelligence system will access any adaptive representational network that has a "physiological sense of being cold" (internal state) component, and match your current environment and stimulus with one from your past that had a successful outcome.

Suppose, for example, you have been playing with your children, throwing snowballs—perhaps something you haven't done since your own childhood—and your hands are now bitterly cold. At this moment, you can access any number of adaptive representational network "movie clips" that represent scenes of a situation in which you were cold, took some action, and shifted your internal state. Essentially, your intelligence system consults your personal menu of networks that have an internal state component of "cold." What fits best? Coming out of the ocean shivering and briskly drying yourself with a towel? Arriving home soaked from a downpour and immediately taking a hot shower or bath? Going to bed and huddling with your spouse for warmth in a power blackout? Walking down a street in Chicago as an icy wind cut through your clothes, and taking shelter inside a doorway? There are as many scenes as there are occasions when you solved the Being Cold problem.

But which behavior will work in *this* situation? Sheltering inside a doorway and huddling in bed are inappropriate. So is the ocean situation. Taking a hot shower certainly might be a solution, but the situation doesn't map very well either. Ah . . . but here's an old movie clip that represents you, age eight, feeling cold after playing in the snow; going inside your house, standing at the sink, and running lukewarm water over your hands (your mother warned you not to hold your hands in front of the roaring fireplace, because the temperature transition would be too abrupt and your fingers would throb); and feeling better. Now your mind has found a match to your current problem—an adaptive representational network that was encoded in a similar environmental setting, when you were in a similar physiological state, experiencing similar sensory stimuli. Like any ARN, this one has another element—a behavioral solution that is dispatched by the system when the other components are associatively activated. And so you go inside and run lukewarm water over your hands.

Over time, as your adaptive representational networks are constructed—scene by scene, episode by episode—you build up an autobiography: a chronology of who you were at various times of your life, and in specific situations. A sense of personal history emerges. And a sense of causality. If you think about it, these functions are inherent in the nature of associative networks—they are being constructed and modified, on-line, over time. The genesis of an adaptive representational network, remember, is a change in your state, which then motivates you to perform some behavior that has an outcome: There is a causal linkage of events. From this behavioral track record, you have a sense of your performance over time—which allows you to form a composite sense of your self that has continuity. Even though your self-representation is being refashioned moment by moment, there are so many common elements from time T1 to T2, T3, T4, and so on that you tend to experience yourself as a stable construct. Again, think of the metaphor of snapshots–or frames–being transformed into the smooth continuity of a movie. Just as projecting film at twenty-four frames per second and videotape/TV images at thirty frames per second tricks our visual systems into the illusion of perceiving continuous motion, so it could be argued that the continuity of the self is a kind of cognitive illusion. But it works. It has utility–just part of the bag of tricks handed to us by natural selection.

Your composite sense of self is richly detailed because it arises from adaptive networks that are intricately connected–not only at the level of associations but also in terms of hierarchical characteristics. For example, elements of a representation–a sensory stimulus, say, or a behavioral outcome (a "feeling")–can be associatively linked to all your other representations of similar stimuli or outcomes. That explains why seeing an object like a doll or a teddy bear can bring to mind similar objects–perhaps the teddy bear you had as a child. That memory can then activate another link–the affective state associated with your teddy bear–and you experience the emotions you felt as a child. And that, in turn, can activate other associated memories.

Suddenly, you are rafting along on your stream of consciousness from a teddy bear to Winnie the Pooh to "hunny" to bees and nectar and the blue flowers and the yellow flowers, to neural substrates to adaptive representations to . . . well, somehow in the writing, this neural excursion became our trip. But you undoubtedly had your own itinerary, skipping from stone to stone along the pathways of your past.

The classic example of this comes from the writing of the French novelist Marcel Proust. (This is a favorite of memory researchers—see Daniel Schacter's *Searching for Memory*.) In *La recherche du temps perdu* (originally translated as *Remembrance of Things Past*), there is a scene in which the narrator, Marcel, visits his mother. She serves tea and small French pastries called "madeleines." Marcel dips the madeleine into the tea—and is immediately transported by what Proust describes as "this all-powerful joy." He is mystified. How could tea and cake produce such a wonderful feeling?

He experiments, trying to repeat the moment with more dips of a madeleine, but the joy quickly attenuates. So he concludes—a smart piece of amateur neuroscience, this—that whatever caused the effect "lies not in the cup but in myself."

And then he has a small epiphany. He realizes that he has experienced this taste before—on Sunday mornings in his childhood when he visited his aunt Léonie in her bedroom, and she treated him to a madeleine dipped in her own teacup. He recalls the inimitable combination of sensory experiences that constituted this moment. He recalls her actions, the words that were spoken, the feelings that were experienced, the pleasure he felt. You can see that calling this a mere memory doesn't do justice to the complexity of the event and fails to reveal the workings of your intelligence system. A memory is not a photograph. It is, in a sense, a revisiting of a moment in your life that captures not only the scene but also the significance of the participants' relationships, their internal states, and the outcome of their behavior. It is the activation of an adaptive representational network.

Your neuronal autobiography, an enormous personal

database, also contains representations of other players in the biological marketplace. The episode of Aunt Léonie and the madeleine, for example, would surely have provoked an entire "archive" of Aunt Léonie representations–some good times, some bad times, but all contributing to Marcel's over-arching view of her and his relationship to her. (How we assign values–positive and negative–to the representations of our relationships with collaborators and competitors is ex-plained in Chapter 5.)

There is nothing arbitrary about the circumstances in which these representations were formed, by the way. They had utility at that moment. Here's another example. Suppose the sensory stimulus is the color red, and your immediate as-sociations are roses, blood, fire engine. Each of the associa-tions has been evoked because of an experience that was unique to you. Perhaps you had a little red Radio Flyer wagon as a toddler, or send red roses on your wedding an-niversary. Maybe you are recalling a blood test or an injury, a childhood ambition to be a firefighter or the heroism of firefighters in New York after the destruction of the World Trade Center. The point is this: Your intelligence system has strict rules of admission for representations. Nothing gets into the club of adaptive representational networks unless it follows the rules: An external stimulus and/or an internal state change provokes a behavior with an outcome that shifts your state, your feelings. *That* is what is memorable. For the most part, these representations and associative networks, these scenes from your past, have tenure in your mind. Although they are relatively easy to modify, they are difficult to dislodge or erase–and you can see why: They are part of the fabric of your life, constituents of your self. Every time you revisit them, you revive them.

If associative networks can be thought of as the horizontal dimension of your self-construct, prototypes are the vertical dimension. A prototype is a higher-order representation, an abstraction of different representations of the same type. Instead of a teddy bear, think of a doll that you might have had as a child. Now imagine all the dolls you have experienced since. In your mental database, as a result of your lifetime experience in the doll dimension, you have a series of representations of dolls. They all possess a quintessential quality of "dollness"—consistent characteristics that unmistakably identify them as dolls. (The list might include a baby face, big, wide-set eyes . . . and so on.) This composite sense of doll is a higher-order representation. A prototype. Doll essence. You can go through the same exercise with apples. Granny Smith, Fuji, Jonathan, Golden Delicious, and Gala are all different types of apple. But they all have an apple essence that defines a higher-order apple abstraction: a prototype (shape, color range, texture, internal seeds, et cetera).

The capacity to form prototypical representations has enormous adaptive utility, for several reasons. Suppose you encounter a novel object in your environment. How do you respond to it? If the new object bears any similarity to objects you have seen before—objects for which you have generated prototypical representations—the most closely associated prototypical representation will be activated automatically as your sensory systems analyze its features. Is this an apple? Well—not exactly. It's about the same size and has a similar skin texture and color. It has a stalk, which suggests that it was suspended from a tree. Cut it open, and there are seeds. Perhaps this is one member of a category we learn to call fruits. And fruits, in the past, have delivered the goods nutritionally. And so, eventually, we learn to recognize pears and avocados and cherimoyas.

What does this have to do with your sense of self? Well–forget fruits for a moment. Is there such a thing as a prototype *you*? Yes. Here's another analogy. Imagine that, at the end of a year, you look at every single document you signed. Credit card purchases, checks, greeting cards, tax returns. Hundreds, perhaps thousands, of your signatures, all slightly different. But it would be possible to create a prototype–to abstract a kind of higher representation–of your signature. Something similar happens when we act as agents in a year's worth of transactions in the biological marketplace. A higher-level abstraction of who we are–our social signature–emerges.

How is this ultimate sense of self built up? In each of the sectors of the marketplace, there are behaviors that are characteristic of you–essential features that imbue every snapshot, every adaptive representational network with *you-ness*. They can be as small as the gestures that you would reliably and consistently use in certain situations. (If you are in public life, these are the behaviors that impersonators will seize upon. Think of early Bushisms–like the squinting and smirking–being used as character cues by Will Ferrell on *Saturday Night Live*.)

Summon up, for example, scenes that feature you as a mother, as an attorney, as a wife. There will be a number of behaviors that are characteristic of you in each of those roles. They form the representational databases from which you can extract higher-order representations–prototypes, as in

the doll example—of your sense of your self as a mother, as
an attorney, as a wife—all arising from your unique historical record of adaptive networks. And because the record is
always changing on-line as time goes by and you have
more experiences, the various prototypes will change. At
any given time, you may have a sense of yourself as a devoted and caring mother—confident in that role, as anyone who shares that sector of the marketplace with you can
easily observe from your behavior. At the same time, you
may feel like an overstretched attorney who is not giving her
best to the practice. A year later, if perhaps you tried to de-
vote more time to work, your self-assessments could reverse.

This kind of activity takes place in all your behavioral
domains. For example, suppose that, over the years, you
have gone dancing regularly. From those experiences, you
have built up a higher-order representation, the prototype
sense of yourself as a dancer. Or scan through your archive
of cycling scenes. From that historical record of cycling behaviors and outcomes emerges a prototype sense of yourself
as a cyclist. These two prototypes can then give rise to an
even higher-order representation: your sense of yourself as
an athlete. And then, finally, at the highest level of organization, there are abstract representations arising from your
sense of yourself as mother, attorney, wife, athlete . . . and so
on, that lead to the creation of your sense of your integrated
self.

## THE BENEFITS OF A SELF

What's the point of having a self-representation? Well, imagine this: What would you do if confronted by a mugger on

a dark, deserted street? You probably know about the so-called fight-or-flight response to an emergency situation. Basically, your energy resources are swiftly reallocated to meet the crisis. Glucose fuel is diverted from nonessential tasks (like digestion) and transported by the blood to the front-line muscles that will be needed for running or combat. All this activity, all this emotional arousal—a racing heart and sweaty palms, for example—is mediated by the sympathetic nervous system (part of the autonomic nervous system that we mentioned earlier). The response to this 911 call is being coordinated by central command neurons in the hypothalamus and brain stem, firing off neural and chemical directives to restore your interior milieu.

Now think back. Hypothalamic nuclei? You already know that the state of activation of your hypothalamic nuclei is a neural snapshot—a representation—of your body's internal state, your internal milieu. You know that it defines the behavioral problem that your intelligence system is designed to solve. Your physiological state is one of fear—as evidenced by the racing heart and sweaty palms. What behavior can shift your internal state? Well, that depends on you—only you—and your sense of self in this hostile environment.

Remember what you've learned about the construction of a sense of self from the fundamental building blocks of adaptive representational networks. Suppose you have a higher-order representation—a prototype—of yourself as an excellent runner. For years, your intelligence system has been encoding representations of you as a premier athlete—someone who has been winning national races. Maybe you'd be tempted to opt for the flight response and show your attacker a clean pair of heels. On the other hand, if you had in mind a prototype of yourself as a karate expert (built up from years of intensive practice, excellence in competition, and the awards you have received), you might fight back.

This is a crucial point. Your behavioral response in this—or any other—situation is a function of the groundwork laid by your intelligence system, accumulating your life experiences, assembling adaptive representational networks, and constructing self-representations. Strictly speaking, your neural/mental database contains no perfectly species-typical knowledge of the world. There is no generic, knee-jerk, fight-or-flight response. There is no set of different evolved modules from which some "selection circuitry" chooses a response. There is only what *you* would do—given your unique history and your intelligence system's estimate of the costs and benefits of the various behaviors available to you—if you experienced an internal-state change that could be described as "fear."

So what is the utility of a self-representation? Simply—and powerfully—this: It allows your mind to make mental models of the likeliest outcome associated with taking one of several potential paths of behavior in decision-making situations. It can mean the difference between life and death. If you're a sprinter, you hit the road. If you're a tenth-dan karate master, you hit the mugger. If you're like us, you hit the deck. (Philosopher Sir Karl Popper suggested that some kind of internal mechanism that allowed us to preview possible actions "permits our hypotheses to die in our stead." The intelligence system we are describing, with its capacity for building self-representations from adaptive representational networks, does the job.)

These neural/mental models of your self make it possible for you to navigate the physical world and negotiate in the social biological market to get your goals met. Your mind has a complex network of representations that can give rise to a menu of self-representations, each emphasizing different features, highlighting different strengths and weaknesses, reflecting different facets of who you are and what talents you possess. Which of the representations of self comes to the fore depends on the behavioral problem you are currently solving.

## YOUR SELF AND OTHERS IN THE MARKETS OF YOUR MIND

Most parents are entranced by their offspring, especially in their early years. Intoxicated by a cocktail of hormones, they uncritically believe that the incidental by-products of their genetic swap meets are destined for greatness. Maybe the next Albert Einstein. Or Julia Roberts. Or Tiger Woods. Swayed by this positive spin from your domestic cheering section, perhaps you had a burgeoning sense of your self as a child of superlative gifts–the smartest, the strongest, the most physically talented, or the most engaging kid on the block. If you had brothers and sisters, rehearsing their own acts in the family vaudeville, maybe you occasionally got social feedback from them that put a dent in your shiny self-representation. An antidote to the doting of your parents. Even so, your home was a small, well-controlled biological market (usually composed of genetically related individuals who had an investment in your success). A place where you were valued.

As you ventured out into the larger, hard-knocks biological market of school and neighborhood, your sense of
self was radically modified as a result of your experiences.
You realized, perhaps, that other children ran faster, debated
better, wrote better essays, were better dancers, had a keener
sense of mathematics, or showed greater fluency with lan-
guages. The pond was bigger, and there were more fish. Your
range of self-representations became more varied as you re-
positioned yourself to find success in the marketplace. Your
behavioral-intelligence system worked overtime to deliver
up-to-date estimates of who you were in relation to the physical and social world. And it used those estimates to make
sound behavioral decisions, steering you along a behavioral
path on which you were likely to succeed.

The challenge, of course, is that all these pathways are
populated by other people. They are part of the equation of
your negotiations in the markets of life. Some of them may
have behavioral strategies that complement yours; others
may be in conflict. So having a model of your self that is
restricted to you alone isn't enough: It must incorporate
your best estimates of all the other important players in your
social landscape. You need to have neural/mental models
that reflect your perception of the characteristics of the people you interact with—their talents, personality, physical
abilities, social positioning, and so on.

How do we figure this out? Partly by personal observation, partly from social data. We can get some of this kind of
information simply by observing another player's movements and behavioral outcomes in the physical and social world over time. (Is there a track record of successes,

or are you dealing with a serial loser?) We can also solicit the perceptions and opinions of other people. (Is the individual liked and respected in the community?)

All of this information updates our representations. But often we need information about other individuals that is not so apparent—information about the contents of their minds: an understanding of what they believe, what knowledge they hold in their minds at any given moment, an understanding of their desires, and what motivates them. To negotiate successfully in the biological market—to cooperate or compete—we need to know what other people know, what they value. We need to be social detectives—mind readers (not in the telepathic sense; rather, being able to explain and predict the behavior of others by putting ourselves in their mental shoes).

Between the ages of three and five, when the cortex is coming into its own and foundational adaptive representations are being constructed, and the earliest representations of self as social agent are making their debut, humans demonstrate a remarkable inferential capacity. We develop the ability to detect whether or not another individual has specific knowledge in his or her mind.

At the most basic level, this ability depends on our visually perceiving, then remembering, what the other individual has been able to see (and, therefore, has probably learned). For example, imagine that a five-year-old boy sees you put a candy bar in a desk drawer in full view of another child—say, a little girl. After the little girl has left the room, the boy sees you take the candy out of the drawer and stick it in your pocket. The boy will know that the little girl still thinks that the candy is in the drawer simply because he knows that she did not see you move it to its new location. He will correctly infer a bit of her knowledge content— a piece of her mind, which is something he could not do when he was a year or two younger.

Cognitive scientists have developed a number of experimental situations like this to understand what a child knows and when. For example, suppose the boy is only three years old, and the same scenario is followed: You switch the location of the candy after the girl has left the room. And now you ask the boy: When she comes back in the room, where will Sally think the candy bar is? The vast majority of three-year-olds will say—in your pocket. They are unable, at that stage of development, to make appropriate inferences about another's behavior. By the time they are four, most can pass this "false belief" test, and five-year-olds, as we said, ace the problem. (Incidentally, there is some evidence that the age is lower for children from large families. Perhaps the more intense pressure to develop behavioral strategies in such a comparatively complex domestic marketplace accelerates a child's ability to mind-read.)

During this developmental period, we begin to infer what someone might be motivated to do, simply by observing what they are paying visual attention to. For example, if we see someone looking at a piece of cake, we infer that he or she is motivated to eat it. Later, when we see two individuals look at some person or object, then exchange glances, we infer that they are thinking about the same person or object, that they have beliefs or desires or intentions. This ability to attribute mental states to other people is our passport to the social world. (Although it is routinely called "theory of mind," we prefer "theory of other minds"–ToOM.)

How do we develop this ability? Is it some inherited mental "module" coming on-line? We don't think so. At first glance, our ToOM abilities may seem to be evolved mechanisms, part of our genetic inheritance. But like all of our other inferential circuitries, they are a natural consequence of the hierarchical properties of adaptive representational networks. What we have inherited is the capacity to *construct* them.

Think back for a moment to what we said about prototypes. Higher-order adaptive representational networks (ARNs)–like "doll essence" or "apple essence"–represent essential information abstracted from the lower-order ARNs that gave rise to them. And when we repeatedly experience the same form of *relationship* between objects and events, we construct higher-order ARNs that represent the essence of that relationship.

Suppose, for example, you are at a birthday party for five-year-olds–relatively recent graduates in the school of mind reading–and the cake has been cut and distributed. Here's one of the possible networks your intelligence system might construct. I'm looking at the cake (stimulus); I'm motivated to eat it (state change); I'm eating it (behavior); I feel good (outcome). At the same time, you're observing the behavior of the other children. Suzy looks at the cake, takes a mouthful, and makes a happy face. Same thing with Kent. And Cathy. And so on. This information is being represented on-line. What does your intelligence system do with this information? It generates higher-order representations. Something like: X is looking at a cake and is probably motivated to eat it. Or: X is looking at a cake and probably *will* eat it. When a situation with similar features arises, these higher-order representational networks are activated and you make an inference about someone else's motivational state; you are able to guess what they might do next. Cathy is looking at the cake. She wants it. You've just read her mind.

These inferencing abilities also underpin our ability to empathize and sympathize with others. (Been there, done that, *felt* that.) Other inferential circuitries, constructed in precisely the same way, support our ability to profit from metaphors and analogies and to grasp abstract concepts. When Shakespeare writes: "Shall I compare thee to a summer's day?" or "How like a winter hath my absence been," we don't imagine that he is delivering a weather report. He is essentially inviting us to activate our ARNs of our experiences of summer and winter. And he is assuming—not a great stretch—that most of his readers have positive individual recollections of summer (blessedly warm, long days, nature in full bloom) and comparatively negative memories of winter (cold, dark, and dormant). The final step, of course, is to assign these evoked abstract attributes back to an individual or situation, which happens as your system is mentally modeling the cognitive problem.

In short, ToOM mechanisms allow us to decipher and describe the behavior of other agents in a biological marketplace. We know—precisely because we do it ourselves—that other players are modeling possible futures, seeking to form alliances, assessing potential competitors. We know that they have behavioral intelligence systems like ours that are computing costs and benefits moment by moment, frame by frame, scene by scene. We know they are forming adaptive representational networks—different from ours, of course, but constructed on the same principles. We know they have self-representations that can change over time. And we know that–like us–they must be mind readers: able to infer beliefs, intentions, desires, and motivations; to have a "theory of other minds."

These are the fundamental skills we all bring to every social negotiation. But complex as it sounds, even that isn't the full story. We are not discrete and isolated individuals. Our actions affect other people in our circle of concern, and they have an impact on us. When we form representations of the key players we interact with, we know that they, in turn, are at the center of their own negotiative nimbus–populated by their family members, friends, loved ones, even principal competitors: anyone who is in important relation to them, either positively or negatively. As a result, our representational profile of any single individual has subsidiary representational profiles of the most important members of *their* social world. Do a little math, and you realize that your mental Rolodex is quite detailed and comprehensive. For example, if you have well-developed first-order ToOM representational profiles of, say, twenty individuals, and maybe a hundred less well-developed first-order profiles, you might actually have encoded several hundred subsidiary representational profiles. In effect, your mind contains a highly populated cartography of an extended biological market: a map that is constantly being updated, enabling you to navigate the social world.

Watch any documentary or read any account of life as a hunter-gatherer in tribes like the !Kung, Ache, or Yanomamö, and you quickly notice the almost total lack of privacy. Life is lived very much in the open; everybody knows everybody else's business—and the relative standings of individual members of the market. In such a relatively compact and closed biological market—perhaps the nearest equivalent in the industrialized world would be a "specialty" market like a school, a company, a club, or a social group (maybe even a small town)—social assessments are crucial data. You spend a great deal of time with people. You have the opportunity of observing their behavior, and the outcomes. You build up a representation of each one, then generate estimates of how you and they rank other individuals in the group on any dimension (loyalty, resources, et cetera).

With enough of these estimates, you can generate a kind of mental topography of the population and have a pretty good idea of who is generally viewed as the most popular in the group, or the most gifted member with respect to a particular talent or skill—all valuable information when you're making market decisions.

## ENERGY-EFFICIENT NEGOTIATIONS: TALK IS CHEAP

Mental mappings of the social world may provide us with critical negotiative information—but negotiation entails communication. All social animals have communication systems, from the pheromonal signals of butterflies and the aerobatics of bees to the mating dances of birds. But human language is unparalleled as an energy-efficient means of communicating complex, abstract information.

Scientists continue to argue about whether the capacity to produce and acquire language reflects an "innate grammar" or whether language is learned. The short answer, of course, is both: Language production and acquisition have both innate and learned components. So the real question is this: Are the universal characteristics of language the product of inherited, language-specific modules—a so-called "language instinct"—or a more general information-processing mechanism?

What are the core characteristics of language? (Our focus here is spoken language—but the same argument applies to sign language or any equivalent communication system.) According to *The American Heritage Dictionary of the English Language,* language is "The aspect of human behavior that involves the use of vocal sounds in meaningful patterns and, when they exist, corresponding written symbols, to form, express, and communicate thoughts and feelings."

If you think about it, the production of this "aspect of human behavior" is equivalent to that of all behaviors generated by our intelligence systems. Language is initiated by an intention or motivation—in other words, an internal state. Internal states initiate the sequencing of behavioral information into meaningful patterns. Language consists of words—symbolic information—sequenced into meaningful patterns. Behavioral information is encoded in ARNs as meaningful relationships between internal states, features of the external world (people, objects, et cetera), behavioral actions or events, and their effects on ourselves and others. Words are arbitrary, but socially agreed upon, associates of these same meaningful relationships.

The meaning inherent in a symbol derives from its association to an object or event, and its relationship to you.

And the meaningful association of objects, events, and relationships is the primary characteristic of ARNs. (*All* behavioral information is retained in adaptive representational networks. It is organized and sequenced into meaningful patterns by the core behavioral-intelligence machinery of the basal ganglia. This is the ARN construction machinery–the analog to the bees' construction system that we described in the last chapter. We describe it in more detail in Chapter 5.)

Language acquisition and interpretation also reflect the functional operation of the core behavioral intelligence system. The input of symbolic information is no different from the input of any other information source from the external world. It's simply a stimulus that your system will respond to.

Suppose you are having a conversation. The input to your system is a data stream: a stream of words (and silences–the spaces between words). For the person talking to you, the words (and the spaces) are intended to convey a specific meaning. The words–cat, sat, mat–activate adaptive representational networks that embody their meanings. In other words, if somebody says "cat" (or you read the letters c-a-t), your intelligence system associatively searches your networks for cat representations. As conversation continues, your inferential circuitries are constantly making best-guess estimates of the meaning, revising the estimate, if necessary, as more information emerges in the data stream of speech. This construction and assessment of possible meanings, like the construction and assessment of any working "mental model," occurs in the planning/modeling circuitries of the prefrontal cortex.

We doubt that psycholinguists searching for an innate, inherited language production and acquisition mechanism will find a more efficient one than the behavioral-intelligence system we describe in this book. And one would be hard-pressed to find a more efficient medium for effecting change in the markets of life than language. In terms of energetic costs, talk *is* cheap.

## MORE THAN ONE YOU ON THE MARKET . . .

Listen to yourself speak to others as you navigate the different sectors of your markets—as you communicate with your mate, your child, your employer. Are you always the same negotiative agent? Do you represent the same biological entity in their lives? Obviously not. Each of us has several distinctive roles in life and somewhat distinctive selves.

Remember Laurence Olivier and the comments of exasperated colleagues who were trying to formulate some single version of him to hold constant in their various biological marketplaces? Here's his biographer again: "there apparently coexist too many different Oliviers for the real one to be found." But perhaps the search was misguided. When people tried to locate the authentic Olivier—the core self—maybe they were looking for something that simply didn't exist. Maybe there never was one Olivier—just as there is no simple *you*. We deploy differently nuanced "selves" like personal emissaries or ambassadors to suit the situation.

As William James puts it in his classic *The Principles of Psychology*: "*a man has as many social selves as there are individuals who recognize him* and carry an image of him in their mind. . . . as the individuals who carry the images fall naturally into classes, we may practically say that he has as many different social selves as there are distinct *groups* of persons about whose opinion he cares. He generally shows a different side of himself to each of these different groups." In other words, a different personality.

# 4

## WHO ARE YOU?

You may think that selves and personalities are exactly the same thing—but they're not. Only you are privy to your sense of self. Your personality, on the other hand, is a complex of behavioral characteristics that outside observers identify as "you." And, chances are, you have more than one. We aren't suggesting that there is a rigid separation between your sense of self and your personality. Perhaps the relationship could best be thought of as a dance, with your sense of self being adjusted on-line in response to feedback about your personality.

Here's how *Webster's New World Dictionary* defines personality: ". . . habitual patterns and qualities of behavior of any individual as expressed by physical and mental activities and attitudes; distinctive individual qualities."

Suppose you draw up a short list of people you regard as knowing you reasonably well. How do you think they would describe the contours of your personality? Do you imagine that their descriptions would map onto each other exactly? Or do you think that the feedback from your parents, children, spouse, siblings, your managers or employees in the workplace, and your college friends would reveal significant differences?

We think they would. ("He's a wonderfully warm human being, caring and sensitive—and funny. People forget his sense of humor." "He can seem really stiff at times. A bit condescending." "My dad's always been there for us. He's a great role model." "I could never understand how someone who apparently set such high moral standards could twist the truth for political advantage." And so on. For more examples, read any biography.)

There are two reasons for this: You will almost certainly have presented yourself in different ways in different sectors of the biological marketplace to get the goods you need in the moment; and the people assessing your personality from the outside are viewing you through the lens of *their* own unique sense of self. They evaluate you with intelligence systems already colored by their personal journeys, by their unavoidably different remembrances of their interactions with you (together with information gleaned from larger circles of acquaintances who also have perceptions of you).

They have each built up a kind of dynamic, multifaceted profile of you—and all their profiles are different. And there's another twist to all of this. Because you are also looking from the inside out—through your own lens, which reflects *your* uniquely individual take on interpersonal events—your perceptions of these same interactions will be different from anyone else's.

Although we all own our selves, our personalities are in a sense public possessions, distributed among the intelligence systems of the players who populate the various sectors of the biological marketplace where we have dealings. In other words, "we" are located, as William James puts it, in all those individuals who carry an image of us in their minds.

In his elegy "In Memory of W.B. Yeats," the poet W.H. Auden conveys the flavor of this idea—and the tension between our individuality and our social nature: the evolutionary ties that bind us to other solitudes.

> *But for him it was his last afternoon as himself,*
> *An afternoon of nurses and rumours;*
> *The provinces of his body revolted,*
> *The squares of his mind were empty,*
> *Silence invaded the suburbs,*
> *The current of his feeling failed; he became his admirers.*

At the nuts-and-bolts level of neurons and blood cells, of oxygen transport and pH levels and action potentials, the fact is that, on January 28, 1939, William Butler Yeats's intelligence system failed. The moment-by-moment reiteration of his self-representations—regular and reliable as the electronic raster that paints updated images on the cathode-ray tube of a TV set—shut down. So far as we know, this was his last moment as himself. And yet the shadings of his personality lived on in the minds of those who had known him. He became his survivors.

We all carry in mind images of loved ones who have died, of former friends we no longer see, of those we have encountered in pleasant or painful circumstances on our journeys through life. This commonplace miracle of evoking past experiences—of re-presenting people, places, and events to our mind's eye (accompanied inevitably by physiological responses)—is an astonishing (but logical) consequence of the facts of life.

Remember the fundamental design features of your intelligence system?

- Behavior is energetically expensive. And so you need an intelligence system that can assess the relative costs and benefits of different behaviors.

- For human beings, one key variable in the cost-benefit equation is our interpretation of the behavior of other human beings. To track that, you need a system that can maintain and modify a database containing information about other players in the marketplace—a history of their relationships and dealings with you (and each other).

- The sweet evolutionary trick that makes this all possible is the adaptive representational network (ARN). These networks form your relational database and make it possible for you to estimate your likelihood of success as you navigate through the uncertainty of the social world.

- New information automatically updates your system in real time.

Now let's look in more detail at the notions of self and personality. First, your sense of self.

Let's agree that William James was right—that we each

deploy different selves to act as our agents in different social situations. This, in turn, is a function of who we need to negotiate with and what we need to negotiate for. These terms in the equation of life are variables, always changing. You play many different roles throughout your life. You are son or daughter, mother or father, brother or sister, spouse, grandparent or grandchild, employer or employee, teacher or student, lover, friend, or enemy.

In each of these different roles, your behavior, your sense of self, is transformed. You are a collection of personas, each identity delicately and strategically nuanced. For each role that you play, your intelligence system draws on a customized database of adaptive representations that, in a sense, shift you into character, all of a piece. It's as if one of the selves in the repertory company that is *you* is waiting in the wings for its cue. And you're on. You walk the walk; you talk the talk. Biochemical systems crank up in, say, mother mode, and you are awash in oxytocin. Or you are in professional mode, you receive a promotion, and your blood serotonin levels are elevated (more on this in Chapter 6).

The key point is this: The representational database associated with each of the roles you play is distinct from the database associated with any other role. The distinctions—a historical accumulation of your experiences in each role—guide your behavior and give rise to the sense of self that is appropriate to each situation. Your behavior—and your sense of self—in each relationship and in each sector of your biological market is custom-made for the goals your intelligence system is attempting to achieve.

If you have followed our argument closely, the chances

are that by now you may be experiencing a kind of cognitive vertigo. After all, in Chapter 3, we noted that a rose is not a rose: A rose is what you make of it. Now, we are suggesting that there is no psychological gold standard—no common currency—for assessing other players in the market of life. We are saying that all of us have individualized databases that create our judgments. Our evolved intelligence systems prompt us to present different selves in different social situations. And everyone else we negotiate with is playing the same game, by the same rules, *but with different data.* No two observers see us in exactly the same way, and our own sense of self is at best an approximation to (and could be radically different from) the way others see us. Strictly speaking, none of us—not even a Supreme Court justice—is capable of making an objective assessment. Everything we do is subjective and intensely relational.

This is exactly how natural selection has arranged things. How could it be otherwise, given our evolutionary history? If we possess intelligence systems "designed" by natural selection to allow us to navigate through an uncertain future—where the primary determinants of our success are the beliefs, the motivations, and the behaviors of other people—the kind of relational system we have described is essential.

Playwrights, actors, and writers—people who create alter egos all in a day's work—have long intuited this. When T.S. Eliot has his character J. Alfred Prufrock preparing "a face to meet the faces that you meet"; when Walt Whitman says, "I am large, I contain multitudes," they are essentially describing life in the marketplace.

In *Orlando: A Biography* (which was actually a work of

fiction), Virginia Woolf wrote about "these selves of which we are built up, one on top of another, as plates are piled on a waiter's hand"; and later remarks that Orlando "had a great variety of selves to call upon, far more than we have been able to find room for, since a biography is considered complete if it merely accounts for six or seven selves, whereas a person may have as many thousand."

Of course, you also have an overarching sense of yourself as an integrated individual. This is a highest-order representation and is basically an abstraction from your various negotiative selves—representations of your behavior in the various roles you play. These role-based representations, in turn, rest on a foundation of all the selves you have been in the past. Woolf's metaphor of plates piled on a waiter's hand is perhaps too stratified and static an image to convey the shimmering dance of representations, but it does underscore the importance of having a stable and coherent foundational sense of self. If the bottom plate is damaged or pulled away, the entire stack may totter and collapse.

We all vary in the bottom-plate department, and that seems to be largely a function of our social experiences in the first few years of our lives. (We're not saying that the stability of your sense of self can't be changed as a result of ongoing experiences in later life—just that it's more difficult.) Because a self-representation is a relational mapping between you and the effects of some other object or individual in the world, consistency is important. The more consistent the social feedback you received as a child, the more coherent and integrated will be your foundational representations of self.

Let's assume that you had a stable childhood. In other words, the biological market in which you first learned to negotiate with others was highly consistent. You would have learned that different behaviors had predictable consequences, and your intelligence system would have encoded this information as a series of foundational representations.

As a result of negotiations with your parents, you would also have learned the boundaries of acceptable behavior, which defined the parameters of your effective social self. If you wanted to get the goods you needed, you had to learn the house rules. You can easily see how this training of your intelligence system would structure and promote the development of a sense of social "right" and "wrong," which mirrored the moral and ethical codes of your parents.

As your circle of concern expanded, your parents would also have established rules for other interactions—with your brothers and sisters, childhood friends and their parents, with teachers, and with strangers. In the process, they were shaping your first self-variations, the first roles in your personal repertory company.

In this structured but supportive environment—where your place and value were well established and rewarded— there would have been very little challenge to your developing sense of self. Perhaps a tremor when your younger sibling was born and you noticed a shift in the flow of parental investment. Maybe another tremor when you first entered school. But nothing could have prepared you for one of life's great seismic shifts: adolescence. If you were like most of us, this was when your sense of self seemed to be in crisis—a crisis of identity.

When you entered adolescence, you crossed the threshold from the developmental life stage into the reproductive life stage. The release of a potent cocktail of hormones, scheduled by the life history regulatory system, established some new and very different goals for you. (As it does at every major life transition; see Chapter 7.) It was a time when you were about to enter the adult biological market-place. Who *you* were in this new life stage was unclear. What were your prospects? Who might you become? How would you be valued in the new market?

In adolescence, your peers began to assume an even larger role in your life than they had during your childhood, when your parents were the dominant influence in your bio-logical marketplace. Your peers began to oust your parents as sculptors of your self, because they were the primary peo-ple with whom you had to compete—or collaborate—for the goods you sought. But does this mean that peer power is such a potent force that it essentially trumps the prior social-izing efforts of parents? Like the tired nature-nurture debate, the peer-parent dichotomy is false and unhelpful.

It is certainly true that your adolescent peers can vote you Most Likely to Succeed, or homecoming queen, or smartest, or most athletic—probably adding a glow to your self-esteem (assuming these are accolades that you value). They can provide a life-altering first sexual experience; you can fall in love. These are undeniably powerful influences.

But it is also true that, if you have ambitions to be the suc-cessor to Venus Williams, Tiger Woods, or Lance Armstrong (assuming you have the physical skills), your coach may be a prime controller of resources in your marketplace. If you want to be a poet or novelist, your English teacher may transform your life. If you want to be a fighter pilot someday, impressing the commandant of your ROTC may be a better investment of your time than dating the star quarterback. And the choice you make in any of these situations will be a function of your customized mind.


## WHO YOU ARE AND WHO YOU THINK *THEY* ARE . . .

We said at the beginning of this chapter that only you have a sense of your self. Other people have their own individualized take on your personality, on you as a social entity in their world. In the quid pro quo of the marketplace, it follows that it is largely your assessment of their personalities that determines whether or not you want to interact with them.

In his novel *The End of the Road,* John Barth suggests that, much of the time, we are only too willing to settle for a quick and woefully incomplete stereotype of a personality.

> As it was, my present feeling . . . was essentially the same feeling one has when a filling-station attendant or a cabdriver launches into his life-story: as a rule, and especially when one is in a hurry or is grouchy, one wishes the man to be nothing more difficult than The Obliging Filling-Station Attendant or The Adroit Cabdriver. These are the essences you have assigned them, at least temporarily, for your own purposes, as a tale-teller makes a man The Handsome Young Poet or The Jealous Old Husband; and while you know very well that no historical human being was ever *just* an Obliging Filling-Station Attendant or a Handsome Young Poet, you are nevertheless prepared to ignore your

man's charming complexities—*must* ignore them, in fact, if you are to get on with the plot, or get things done according to schedule.

Categorizing people like this is hugely tempting. We reduce the problem-solving space by assigning identifiable roles—stereotypes. It is no accident that roles are usually written for actors to convey the personality of their characters as quickly and economically as possible. Take a look at any soap opera: We may no longer wear the identifying masks from Greek theater, but it's easy to spot the characters the actors are portraying (The Handsome Young Poet or The Jealous Old Husband, for example).

Inevitably, getting to know people—forming a multi-dimensional profile of them—is a much more complex process. As we begin to play this social chess, our intelligence systems construct adaptive representations that allow us to create mental profiles of individuals and their worlds and predict, with varying degrees of success, the way they might behave in the future. Over time, with enough case studies to provide a database, we form higher-order representations of the behavioral characteristics that seem to reliably predict outcomes over a population of the people we encounter in the marketplace.

We are not suggesting that we have some kind of "personality perception system," with genetically predetermined neural circuits—so-called modules of the mind (see Chapter 8)—that specialize in alerting us to such qualities as honesty or loyalty. A personality, remember, is essentially a bag of tricks for negotiating in the biological marketplace. The assessments we make of other players are individualized and responsive to new information in real time. And different people use different criteria for making assessments, depending upon what they place most value on in any given situation.

Think about it: How do *you* assess the value of a potential partner—or competitor? What yardsticks do you use? Physical attractiveness? Emotional compatibility? Good theory of other minds (can reliably intuit other people's intentions, beliefs, motivations)? Sense of humor? Adventurous, outside-the-box, personal style? Are you looking for a business partner or a partner for life? If you're looking for a business partner, what business are you in? Selling cars, selling weapons, or selling spirituality? It makes a difference. And if you're looking for a mate, to what extent is your personal "radar" calibrated by your unique historical experience of your family dynamics? Your last date?

Our perspective is significantly different from some of the standard folk taxonomies that psychologists use. But there are some correlations. Take as an example a widely used classification system that partitions personality traits into the so-called Big Five categories: extroversion; openness to experience; agreeableness; conscientiousness; and neuroticism. We don't think that negotiators assess each other in only five dimensions—but let's explore them and see how they map onto our view.

*Webster's* defines extroversion as "an attitude in which one's interest is directed to things outside oneself and to other persons rather than to oneself or one's experiences." Sociability and gregariousness are often used in the same context. In other words, extroversion is the extent to which a person is actively participating in the activity of the outside world. Knowing how open someone is to experience helps us to predict how likely they would be to try new ventures, new relationships, new techniques, and new ways of thinking about things. Now, whether *you* value openness depends on your own makeup and your needs—but one way or another, knowing where someone lies on this dimension is important information if you are considering a partnership of any kind.

Or take agreeableness, variously defined as cooperative, "conformable or in accord," "ready to consent or submit." This is important information to have about a potential cooperator or competitor—for obvious reasons. The conscientiousness dimension assesses traits like persistence and dependability: the degree to which someone's behavior appears to be guided by their conscience rather than circumstances. Having a sense of this allows us to predict how loyal they are likely to be and how they will follow through on a project in a principled fashion regardless of unforeseen challenges arising in an uncertain future.

Neuroticism is a measure of emotional instability and maladjustment; it is obviously useful to know how prone someone is to behaving in negative and abnormal ways if you are expecting to interact with them at some point. Would you voluntarily work with someone you knew to be emotionally unstable, prone to depression? No? If you are a bank manager, or a member of the transition team that vets candidates for Cabinet posts, that's probably a wise choice. But suppose you are dealing with an actor, or a rock musician? Or—courtesy of a time machine—suppose you received a manuscript titled *Orlando: A Biography* from a writer called

Virginia Woolf, an incest survivor who had four breakdowns before she was thirty?

Some people—like Woolf, who eventually committed suicide—seem to have more difficulty navigating the social world and negotiating in the biological market than others. Maybe this has happened to you—even in a limited fashion: an episode of depression, perhaps; a sleep disorder; a pervasive sense of anxiety. If you have ever taken your troubles to a psychologist or psychiatrist, you probably know that your therapist will classify your problem according to a taxonomy devised by the American Psychiatric Association: the *Diagnostic and Statistical Manual of Mental Disorders* (or *DSM*). All told, there are almost four hundred of these categories and subcategories. Get a ticket from the personality police for a 295.30, and you have been diagnosed a paranoid schizophrenic; a 309.81 signifies post-traumatic stress disorder; and so on. The vast majority of these codes describe a "disorder," which, by definition, is an upset of normal functioning.

The current edition (*DSM-IV*) lists eleven "personality disorders"—patterns of social behavior that are generally held to be ineffective, counterproductive, or just plain difficult for the rest of us to deal with. 301.0, for example, is Paranoid Personality Disorder; 301.6 is Dependent Personality Disorder; and 301.4 is Obsessive-Compulsive Personality Disorder.

Let's focus on just one of the eleven—Dependent Personality Disorder. To qualify as a card-carrying 301.6, you need to demonstrate "a pervasive and excessive need to be taken care of that leads to submissive and clinging behavior and fears of separation." You can do this by achieving a passing grade on at least five of the following requirements:

- You have difficulty making everyday decisions without an excessive amount of advice and reassurance from others.

- You need others to assume responsibility for most areas of your life.

- You have a hard time expressing disagreement with others because of fear of loss of support or approval.

- You have difficulty initiating projects or working alone (more from a lack of self-confidence than a lack of motivation).

- You go to excessive lengths to obtain nurturance from others (even volunteering to do things that are unpleasant).

- You feel uncomfortable or helpless when alone.

- You view the ending of a close relationship as a catastrophe, and urgently seek a replacement as a source of care and support.

- You are unrealistically preoccupied with fears of being left to fend for yourself.

The clear message here—according to the *DSM-IV*—is that individuals diagnosed as "dependent personalities" are too passive. They allow someone else to direct the course of their lives. As part of the unspoken contract, they subordinate their own needs. It's hardly surprising that dependent personalities are seen as lacking in self-confidence.

But here's a different way of analyzing the situation.

Could this kind of personality profile be *functionally effective* in the biological marketplace? Is it possible that being a dependent personality (DP) is just another way of getting the goods? Think about it. The profile describes an individual who has—for whatever reason—entered into a very specific negotiative agreement.

According to the terms of the deal, the DP doesn't have to expend any time or energy making decisions, doesn't have to assume any responsibility for the consequences of the outcomes of decisions, and doesn't have to pay the inevitable "costs" (as well as avoiding the risks) of negotiating in the market at large. But there's a penalty: The DP does pay the "cost" of subordinating his or her individual desires to those of the partner.

Take a look back at the *DSM-IV* personality profile, and you'll see that this kind of relationship *requires* that one of the individuals meets the major behavioral criteria for a "dependent personality disorder." Is this an arrangement that would work for you? Maybe not—but there is no reason to think that it couldn't work perfectly well for some people. Which raises this question: If the dependent individual is in the relationship willingly, and his or her major life goals are being met, is there any justification for considering such a person to be dysfunctional or mentally "disordered"?

What about the claim that dependent personalities suffer from low self-confidence? Is that a symptom of disorder, or does it make adaptive sense in our framework? Your level of self-confidence, remember, arises from your mental database of your behavioral outcomes. Professional sportsmen can consult statistics in the record books—how many RBIs, how many successful free throws, and so on. For the rest of us, the records of our times at bat in the social world are kept by our intelligence systems: We have what you might think of as an inner statistician. (The inner statisticians of the people we interact with are also keeping tally, of course—and may not be so charitable about our performance.)

To compute a kind of "confidence index," your inner statistician calculates the ratio of successful outcomes to attempts. But because dependent personalities have comparatively limited experience as negotiators in the biological market at large—very few times at bat—the DP inner statistician has only limited data.

The functional utility of having an index of self-confidence is that it serves as an estimator of how you will make out *next* time at bat in the market. Based on past experience, it predicts the likelihood of success or failure of alternative behavioral options that your system is evaluating at any given moment.

But what happens if your database is limited? The intelligence system shifts to a default level of confidence that is low, so that you proceed cautiously. If you have no training as a sushi chef, you would be wise to avoid the blowfish. And if you have limited experience in navigating the complexities of the biological marketplace, the same argument applies. Ill-founded excesses of confidence can get you into trouble. For a dependent personality, then, it could be argued that having low self-confidence is a protective feature: functional rather than *dys*functional.

How is a dependent personality created? If you think about the principles governing the construction of adaptive representational networks (and how they regulate behavior), all you need is a small, closed biological market in which virtually all of the goods are regulated by someone else. Sound familiar? Something like the familial market in which most children are raised. This is the nursery of dependence—the earliest, foundational setting where a personality might be skewed into a DP style of negotiation. (We are not saying, obviously, that all children are transformed by their parents into enduring dependent personalities—simply that the potential exists most powerfully in this setting.)

As a child, you might not even have registered the fact that virtually all of your decisions were being made for you. The fact that getting your needs met involved a negotiation—or, more likely, a downright capitulation ("If you don't do A, you don't get B")—probably seemed like business as usual. But even if you did notice (and didn't like it), there was probably precious little that you could do about it. The paradoxical fact is that in childhood—the environment in which a dependent personality is most likely to develop—it can be the most functionally effective personality to have.

But transport that personality across market lines, and you can be in trouble, because you are now trying to make independent life decisions with a deficient database (and a confidence index that is set to a default of *low*). And so the chances are that you would be predisposed to get involved with someone who wanted to dominate you—to control your access to the goods you need. This, after all, is what your intelligence system has been conditioned to respond to. But the likelihood of finding a dominant partner who would care for your basic health and well-being the way your parents did would be low. It's not surprising that dependent personalities are at a higher-than-normal risk of becoming victims of abuse.

What are the lessons from this kind of informed analysis?

First, personality characteristics, like other behavioral characteristics, develop largely as a function of our experiences. This is a consequence of the fact that our intelligence systems are designed to enable us to get our life goals met in the social environment we are navigating through at any given moment. This is normal behavior.

Have you ever felt unwilling to get involved with people unless certain of being liked? Have you on occasion been preoccupied with being criticized or rejected in social situations?

How about exaggerating your achievements and talents? Have you ever had the sense that you are special and unique and can be understood only by other special people? Maybe you've been envious of other people, shown arrogant behavior?

Are you uncomfortable in situations in which you are not the center of attention? Have you ever used your physical appearance to draw attention to yourself? Maybe demonstrated a certain theatricality, an exaggerated expression of emotion?

If you answered yes to all of the above, you have some of the defining characteristics of 301.82–Avoidant Personality Disorder, 301.81–Narcissistic Personality Disorder, and 301.50–Histrionic Personality Disorder. Does that mean you have an illness? Probably not. We suspect that very few of us could probably read through the Personality Disorders section of the *DSM-IV* without feeling that we had, at some time in our lives, been there, done that, and felt that.

Does that mean that we were suffering from a "disorder" of the mind? We think it is more constructive to put it this way: Rather than being *dis*orders, these behaviors are actually orderings of our adaptive representational networks, providing solutions to social negotiative problems that we once faced. But the fact that they were appropriate then doesn't mean that they are appropriate now.

And that is the second point: Change is possible. While some aspects of what we consider to be someone's personality, such as their temperament, might have a stronger genetic component than others, all aspects of the personality are shaped by experience and are subject to change.

A NEW YOU?

Your behavioral intelligence system continues to adapt online as a function of ongoing experience. That means it is possible to modify behavioral patterns—even the most ingrained personality characteristics. But it isn't easy. It takes setting up appropriate environmental contingencies, which is not always possible. And it requires many efforts over long periods of time—particularly if you are trying to modify personality profiles that involve strong emotional underpinnings, like the paranoid profile, which is characterized by strong fear and anxiety components. Why? Because experiences that induce fear produce long-lasting and generalized effects—they are designed to make us err on the side of safety in the future.

Even so, behavior is always being modified by our intelligence system's adaptive mechanisms. Every time there is a successful behavioral outcome, the adaptive representation that produced the behavior gets stronger; every time the outcome is unsuccessful, the representation gets weaker. The sheer cumulative power of repeated experiences is rather like etching behaviors deeper into your psyche, making them more resistant to change.

Just think how many times during a single day you activate behaviors that are characteristic of your usual personality style. Then figure out how long you have been constantly reiterating those behaviors over the course of your lifetime. That will give you some idea of the kind of effort it would take to refurbish this personality—the number of times that an alternative set of behaviors would have to be successfully used before the changes that were occurring on the neural level would be observable on the behavioral level. Remember, too, that your foundational personality, like your foundational self, remains the baseline for change. New experiences can affect only the course of ongoing modifications of your intelligence system. And it takes a concerted effort over a very long time to produce observable change.

There's another reason why personality change is so difficult: other people. As we said earlier, your personality is public property, distributed in various incarnations in all the individuals who carry an image of you in their minds. You have a negotiative history with all these people. Every time you produce a behavior, they deliver what they imagine is an appropriate social response. They then add the salient details of the transaction to their theory of your mind—their version of your personality. Obviously, the closer and longer your relationship, the more powerfully this personality will be etched in their minds.

Now suppose you want to effect a positive change—to

make some ongoing but enduring shift in the behaviors that other people identify as characteristic of you. For a new behavior to create an enduring representation in your mind, it has to reliably produce a successful outcome. In this case, success is measured by an appropriate social response. And here's the catch: The people who could reward you with that response have a history of interacting with you in a fashion that was largely shaped by the old you. Because change is as difficult for them as it is for you, it is an uphill battle (even if they were motivated) to edit the image they have of you and your behavior. And so they are inclined to react to you as they always have—which means that you don't get the feedback you need to lock in the new behavior.

A related problem is that we are simply resistant to recognizing or acknowledging change in others. Our minds are designed to attempt to predict the future: We form impressions of other people that enable us to mentally model an estimate of their future behavior. These impressions—an attempt to hold static something that is inherently in flux—are highly resistant to change. That's why first impressions can be so powerful and long-lasting. And because our sense of self is a relational construct, a failure on the part of the social world to acknowledge and mirror back a change in your personality and underlying self undermines your efforts.

Given these barriers to change, how can you shift? "Just give up the state of mind you're in," suggests the pop group Sister Hazel in a recent song. "If you're tired of fighting battles with yourself/If you want to be somebody else/Change your mind."

It would be more accurate—if less lyrical—to say: If you

want to be somebody else, change your marketplace. In other words, don't let people whose representations of you were formed under the regime of the *old* you be your primary source of feedback. If you are trying to establish a new behavior, it may be easier to achieve in a marketplace populated with cooperators who have no investment in the way you were. Many "step programs," such as Alcoholics Anonymous, embody this approach and seem to have intuited these deep principles of the human mind.

Programs like AA (and the many nonreligious versions) instantly supply you with a new marketplace, with a new set of cooperators, a new set of contingencies and rewards. This is an environment where your new behavior (in this case, sobriety) produces a reinforcing social response. Over time, new adaptive representational networks are created that may prove more potent than those that once held you captive. Your level of self-esteem improves. How does that work? It's a story of neurotransmitters and the control they exert over your behavioral intelligence system—a story we tell in the next chapter.

# 5

## ANSWERING THE
## CATERPILLAR'S QUESTION
*What Size Do You Want to Be?*

On July 4, 1862, a mathematics lecturer at Christ Church, Oxford University, took the three young daughters of the college's dean, Henry Liddell, on a boating trip. To while away the time and amuse the children on a lazy summer afternoon, he improvised a story. It began with Liddell's ten-year-old daughter, Alice, falling into a rabbit hole and having a series of adventures underground, where she encountered a motley crew of characters, including a hyperkinetic White Rabbit, a sociopathic Red Queen, a troupe of dancing lobsters—and a Caterpillar, perched atop a blue mushroom, smoking what would probably now be described as a controlled substance through one of those long oriental pipes called a hookah.

The mathematics don was Charles Lutwidge Dodgson. We know him better by his pseudonym: Lewis Carroll. Three years after the boating trip, his yarn was published as *Alice's Adventures in Wonderland*. And curiouser and curiouser, as

Alice might have said, Carroll's text offers us a metaphor for this chapter: the way your behavioral intelligence system is calibrated by three monoamine neurotransmitters—serotonin, dopamine, and noradrenaline. (You can think of neurotransmitters, incidentally, as messenger molecules that chemically ferry information from cell to cell in your brain, across gaps called synapses. The word *synapse* comes from Greek words that mean to join together: a junction.)

But let's catch up with Alice, hotly pursuing the White Rabbit. She quickly discovers that, to navigate through each environment, she needs to change her size. Sometimes, she is far too big to get through a door; other times, she is too small to reach the handle. Her stretching and shrinking—she thinks of herself as a collapsible telescope being opened and closed—are mostly mediated by the substances she ingests. First, there's a bottle labeled DRINK ME—which ratchets her down to ten inches tall. Then she finds a cake marked EAT ME—and shoots up over nine feet tall.

By the time of her surreal encounter with the Caterpillar, Alice is three inches high. When the Caterpillar demands to know who she is, she is stuck for an answer. "I know who I was when I got up this morning," she tells the Caterpillar, "but I think I must have been changed several times since then." I'm not myself, says Alice: "being so many different sizes in a day is very confusing." And so the Caterpillar asks: What size do you want to be?

Answering the Caterpillar's question—and several others that he would never have thought of—is a fundamental goal of this chapter: to explain how monoamines globally calibrate your intelligence system. We'll start with serotonin. Our answer to the Caterpillar's first question is this: You want to be the right size for getting the goods you need. If this seems, at first sight, as obscure and enigmatic as the sayings of the Mad Hatter, just think of size as a metaphor for status—more precisely, for the place you think you occupy in a particular biological marketplace. What kind of self-representation will best achieve your goals? What range of social behaviors is appropriate for the situation you are in?

Your sense of self-worth—your self-esteem—shifts throughout your life. In fact, it can shift dramatically during the course of a single day. At any given moment, you are like Alice, attempting to reconcile your sense of self with the situation you are in. Should you be bigger or smaller? What does the marketplace call for?

If you didn't do your homework last night, do you go out of your way to attract attention to yourself in class, or maintain a low profile? If you have been called on the carpet by people with the power to fire you, will it serve you best to be confident and combative, or deferential and conciliatory? Perhaps neither. Maybe being forthright but firm, taking responsibility for any failings but reminding them of your previous good record? Plainly, it depends on your reading of the situation. Where does that information come from, and how do you select the appropriate behavior—the appropriately sized self? (Although we are using size as a metaphor for positioning in the marketplace, other animals make explicit shifts in their apparent size in an effort to achieve their goals. Piloerection—the standing on end of body hair or fur—makes animals appear bigger. And in submissive postures, they appear to shrink, diminishing their height and stature.)

You already know the broad outlines of an answer from

the earlier chapters. Because of the associative properties of adaptive representational networks, your intelligence system can search for information in several dimensions—an environmental stimulus, an internal state change, an emotional response, your behavior and its outcome. Rather like entering a word into a search engine and then hitting the "Go" button, your system can access the salient sets of representations (those movie clips we mentioned earlier).

Suppose you're feeling hungry: Your stomach is rumbling; it's way past noon, and you didn't have breakfast. Your blood sugar is low, and you are getting cranky. These internal state changes are rather like entering "hungry" into your personal search engine. Your intelligence system sorts through your ARNs for scenes where you were hungry in the past. Scenes that have an internal state component of being hungry, and a behavioral outcome that represents "sated" (problem solved—remember the grizzly?). Which of the solutions best matches the current stimulus, the situation you are now in? (Head for the kitchen? Order room service? Drive to a restaurant? Forage for berries? Fashion a makeshift fishing line from vines and a wire coat hanger salvaged from your plane crash on a remote island?) Out of this matching exercise emerges a behavior—the action you will now take to locate food.

Does the system always work impeccably? Do you always have an identical situation in your database—a perfect match—so that your behavior is essentially just a matter of finding the appropriate ARN? Of course not. There are no perfect matches. Situations change. You change. Every behavioral decision is novel.

Suppose you are Tiger Woods, sizing up a ten-foot putt at the eighteenth hole at St. Andrews that might win you the British Open. You have never been faced with this exact situation before. You have sunk hundreds—thousands—of ten-foot putts. But not at St. Andrews. Not at the eighteenth hole. And not to decide a championship. In other words, there is no perfect match in your database of adaptive representational networks linking situation, stimulus, behavior, and behavioral outcome. So what do you do? Recall what we said in Chapter 3. You have representations of yourself as a world-class athlete, as a golfer, as someone who routinely sinks ten-foot putts—plus the crucial information contained in those representations. Are you more likely to make the shot than someone who has no internalized track record of success? Sure—you have what might be thought of as "been near there, done something like that, felt something like that" neural networks. It's not a facsimile, but it's a good working model. And so you make the stroke, and—more likely than not—the ball rolls into the cup.

There is a story told about Laurence Olivier. After a particularly electrifying performance as Othello, his fellow cast members—recognizing that he had surpassed even his usual high standards on this occasion—applauded him off the stage. Olivier, instead of basking in the glory of the moment, stalked off stage and slammed his dressing-room door. One of the other leading actors was delegated to find out why he was so distressed, and called through the keyhole: "What's the matter, Larry? It was great!"

"I know it was great, dammit, but I don't know how I did it," Olivier replied. "So how can I be sure of doing it again?"

The answer is: He couldn't. Nor could Tiger Woods. Nor can you, when you reach some particularly sweet resolution of a novel problem.

That particular conjunction of stimulus situation, internal state, behavior, and outcome may never again be precisely repeated with the regularity of the tides or the predictability of a solar eclipse. (Other actors will deliver their lines in a slightly different way, with different emotional nuances and intonations, provoking different responses. The condition of the eighteenth green at St. Andrews will be slightly different: The wind speed and direction will vary.) But what you can be sure of is that your intelligence system has a record of what worked: You're never starting again from scratch. Integrated into the very fiber of your being is knowledge of what it is like to be at the top of your game–an indelible personal record to which you can now aspire in the future.

Of course, there will be times when you will have a dearth of ARNs upon which to draw for guidance. As we mentioned in the last chapter, if your database is limited, your intelligence system shifts to a low level of confidence, so that you are more inclined to make prudent decisions. Your sense of self will reflect your inexperience: You may feel apprehensive. (First date? First attempt at your driving test? First skydive?)

In terms of our Alice metaphor, your evolved system is essentially counseling you to err on the side of caution and appear too small, rather than too big. Being the right size in the right situation is a matter of survival. Natural selection designed your behavioral intelligence system to associatively ratchet up and down levels of self-representations, moving you from higher to lower (or lower to higher) senses of self as a function of your immediate past history of successes or failures in a specific arena of the social world.

Your system is conservative. It is weighted to use strategies that have a proven cost/benefit track record. In effect, it says: Look, you were doing just fine when you were three feet tall. You were part of a tightly knit biological market. You knew your place. Now you want to play in a different league. Is that wise? Why not stick with the self-representation that matches your current social station? You have a reliable database for being three feet tall. Why stray from your primary market and run the risk of being hurt?

Although it might seem counterproductive to have an intelligence system that keeps you in your place, social stratification—hierarchies and pecking orders (and the modern version, class)—has a logic. Renegades who buck the system can lose their social privileges. In ancestral primate societies, they could be ostracized or die. Knowing your place could be a passport to survival. But how would you know your place? How do you know the answer to the Caterpillar's question: What size do you want to be? The short answer is chemistry: serotonin.

If you know the name at all, it's probably because you've heard of Prozac, or Zoloft, or Paxil. And if you've heard of them, it's probably because you saw some news report or read a book that suggested that a drug like Prozac could boost self-confidence. The mechanism for this epiphany is the increased concentration of serotonin molecules in the biochemical no-man's-land between your nerve cells. Selective serotonin re-uptake inhibitors (or SSRIs) such as Prozac effectively flood the space between neurons with neurotransmitters that streamline the passing on of messages.

What does this have to do with social status? Well, forget humans and Prozac Nation for a moment. Think about vervet monkeys.

Vervets—native to sub-Saharan Africa, with colonies in the West Indies—live in close-knit troops, each with a well-delineated hierarchy. There's always one dominant (or alpha) male and one dominant female. An alpha male gets to be troop president by defeating the other males. Life at the top has certain rewards, including preferential access to limited resources, like females and food, and the best chance of leveraging your genes into the future. Study a vervet troop for a while, and you could quickly identify the alpha male by his behavior and "temperament." He's the laid-back, socially confident monkey, sitting in the best seat in the house, eating the best food, and having the most sex. All the other males treat him with deference and respect. There's a good reason for behaving in a submissive manner to the alpha male (an adaptive utility, to use the technical term): You don't incur his displeasure and run the risk of a severe drubbing.

The same survival logic applies throughout the troop: All the members are submissive to both the alpha and to the other vervets that have higher social status than themselves. Stepping out of line can be dangerous. To act as the equal or superior of a troop member who has repeatedly proven himself dominant to you—say, by approaching his mate, or helping yourself to his food—is likely to provoke an attack. The biological market is a highly competitive arena, and it is vitally important for a monkey to know its place in the dominance hierarchy.

Most vervet troops settle into relatively stable social

hierarchies, but there are always upheavals, and each member's position changes over time. For example, a female is a life member of the group into which she was born (it's called her natal group). But a young adult male will leave home and migrate to a neighboring group. Although his status was once well defined, he now has to establish a position in the hierarchy of the new troop. (Human parallels spring to mind: the new kid on the block, the new boy in class, transfer to a new job—all situations that call for finely honed social skills to avoid conflict.)

And, inevitably, there are the occasions when the alpha male must confront the heir apparent—the contender who was perhaps an immigrant from another group has matured and now aspires to the vervet throne. If the dominant male loses, alliances shift, and a new hierarchy is eventually established.

But what happens in a vervet's intelligence system—its mind—to mediate these shifts in the hierarchy? How does it weather changes in the social climate and establish a new and appropriate social status? How does it know—as the Caterpillar asked—what size to be?

The answer lies in a neurochemical detective story. If you tap into the cerebrospinal fluid (CSF) of a vervet, you can find evidence of what happens in its system when it confronts these kinds of status issues. Here's the major clue. The CSF level of 5-Hydroxyindoleacetic acid (5-HIAA) is elevated in dominant vervets. 5-HIAA is a metabolite—a chemical breakdown product—of serotonin. Measure 5-HIAA levels in CSF, and you have a pretty good measure of serotonin levels in the brains of vervets. The result: Alpha males have about twice the serotonin levels of subordinates, and there is a gradient of levels down to the most subordinate member of the troop.

Now, suppose you remove the alpha male from the troop. You're left with a collection of betas, gammas, deltas, and so on. Who takes up the mantle of leader? The obvious candidate is the senior subordinate—the beta male, the vice presidential vervet most likely to fire up his serotonergic systems now that the daunting presence of the dominant male has been sidelined. But even this succession can be trumped by chemistry. Take any of the other subordinates at random, and inject him with a drug—such as Prozac—that boosts serotonin. What happens? He becomes dominant. His serotonin levels increase until they match the levels of the former alpha (whose own levels eventually fall to those of a subordinate). CSF 5-HIAA levels are a good predictor of a vervet's status.

And yet there's more to this than just a biochemical shot in the arm. It's a two-stage process. First, increased serotonin promotes affiliative behavior. The treated vervet forms a coalition with powerful females, who endorse his candidacy. There is a difference between being domineering and being dominant. Low-status males (low-serotonin males) are on a short fuse, aggressively pick fights (although, notably, not with the alpha male: Instead, they tend to attack juveniles and sometimes females—always a sign of a low-status male), and are less likely to forge cooperative alliances. Dominant (high serotonin) males are confident, relaxed, and skilled politicians—consensus-builders. (Although researchers have always noted the importance of female vervets in the status game, there is little information about female serotonin levels—beyond the observation that they are generally higher, on

average, than males'. But here is an interesting fact: If all
the females are removed from a vervet troop, the serotonin
level/status effects we have described apparently vanish. If
you think about it, this makes perfect sense. The removal of
females removes the ultimate raison d'être for a hierarchy.
Bragging rights and preferential access to procreative partners don't matter if there are no partners to impress with
your manly virtues.)

In a sense, downsizing your aspirations, keeping your
head low and avoiding conflict with those who outrank you,
may keep you out of harm's way. Your intelligence system is
constantly searching for an appropriate self-representation
that will maintain your viability in these adverse circumstances.

Remember the four basic components of ARNs? An
environmental stimulus, an internal state, a behavior, and its
outcome. If you are bested by an alpha male, your internal state changes. Metaphorically, your tail is between your
legs. Your intelligence system recalibrates. Serotonin levels
go down. Now your system is searching for networks associated with less-dominant behaviors—behaviors that had
adaptive outcomes in similar circumstances. Even if you
have led a charmed adult life, where everything you touched
turned to gold and adversity is a novel experience—which
is most unlikely—your system still has an archive of scenes
representing "defeat" from your childhood. The times
when you were grounded by your parents. The times when
a botched interception on your part gave the other side the
victory. The times when your grade on a test was substan-
dard. The times, perhaps, when you backed down from a
playground scrap. Or fought, and were beaten.

Associated with each of these situations is a behavior. Some worked, some didn't. Most likely, the ones that worked involved the eating of a little humble pie, an acknowledgment of a decline (even temporary) in status. Your self-representation changed: On the Alice scale, you became smaller for a while. And so you renewed your efforts on the homework assignment, practiced harder, followed the rules of the house, pragmatically acted in a less confrontational or more deferential manner to the alpha male of the sixth grade, and so on, until you rehabilitated yourself and began to get bigger again.

Vervets, of course, are more concerned about a repeat thrashing by the dominant male than about turning in improved monkey homework. For a subordinate vervet, the raw facts are simple: To survive, it needs access to resources—a square meal in a safe environment. The most available resources it can draw upon are those in its biological marketplace: the social world of the troop. If it migrates to another troop, it may face an even more difficult struggle. And so the problem of defeat is solved by behaving submissively. The vervet's intelligence system selects for self-representations that reflect a lowered sense of social status.

You might say that the vervet has suffered a blow to its self-esteem. Now suppose that this internal state becomes chronic: The monkey is condemned by circumstances to occupy for the foreseeable future this low rung on the social ladder. And let's say that physicians are asked to diagnose its condition. Chances are they might suggest (with appropriate caveats about anthropomorphism—attributing human characteristics to another animal) that the vervet seemed "depressed." Press them a little further, and they might say that the depression was a result of a "chemical imbalance"–specifically, an abnormally low level of CSF 5-HIAA, or serotonin.

Is this a satisfactory explanation? Or does it miss the real point? It implies that the vervet's behavioral intelligence system is broken–dysfunctional. But what could be *more* functional than having a system that ratcheted down your level of self-esteem so that you behaved appropriately to ensure your survival? What could be more functional than having a system that adjusted your social status, on-line, and kept you as a viable (albeit second-string) player in the biological market of the troop?

In other words, perhaps this type of low-level "depression" might be better thought of not so much as an illness but as a path devised over evolutionary time by your intelligence system that enables you to reposition yourself in the marketplace and maintain your viability.

In humans, as in vervets, a diminished sense of self-worth is a symptom of a reversal in the biological marketplace–a manifestation of a functional behavioral intelligence system, not a dysfunctional, chemically imbalanced one.

Of course, there's more to a major clinical depression than the serotonin big/small element. It can involve restlessness and agitation, disturbed sleeping and eating patterns, and in severe cases, a complete loss of motivation and psychomotor retardation. These symptoms result from the changing activational levels in two other monoamine neurotransmitter systems that, like serotonin, also exert ongoing regulatory control over the behavioral intelligence system: the dopaminergic and noradrenergic systems. As with the serotonin regulatory system, the dopamine and noradrenaline systems are designed to help you generate behavior that will keep you viable, even if you are down on your luck and have slipped several rungs on the social ladder.

## DOPAMINE SYSTEMS

No matter what your current position in the social world, you continue to think and to behave. What you think and how you behave are affected by the level of serotonergic activity in your brain. But the fact that you can think and behave at all is largely the result of the activity of another monoaminergic neurotransmitter system: the dopamine system.

We met the insect equivalent of the dopamine system in Chapter 2, when we discussed the honeybee. Remember that the bee has a neuronal system that constructs ARNs and communicates with its life history regulatory system (which monitors its bioenergetic state) to guide its behavior in the environment.

We told you that a special neuron (called VUMmx1) acts as the go-between, projecting throughout the bee brain. The neuron releases the neurotransmitter octopamine, which is a close cousin of dopamine and does much the same job. The human intelligence system has an analog to this arrangement, although much more complex. Instead of just a single neuron anchoring the system, we have a collection of nuclei called the basal ganglia, located deep below our cortical hemispheres in the ancient basement of the brain.

The basal ganglia communicate with the hypothalamus, the central component of the LHRS, and they give rise to three distinct dopamine pathways—the *mesolimbic, nigrostri-*

*atal,* and *mesocortical dopamine systems.* (Don't be daunted by the terminology, incidentally. It's just the neuroscientific equivalent of saying "the Pasadena Freeway" or "the Long Island Expressway." It simply specifies the starting point and the destination of traffic in the brain. *Meso* means "midbrain," for example, so a *mesocortical* pathway conveys information from the midbrain to the cortex.) These dopamine pathways play critical roles in the construction, modification, and activation of the ARNs that guide human behavior and thought. In effect, they are the central machinery of our behavioral intelligence systems—and our minds.

The mesolimbic (midbrain to limbic system) dopamine pathway is responsible for reinforcing behaviors that historically resulted in your getting the "goods." Suppose you are hungry. You could imagine any number of behavioral sequences that would solve the hunger problem. (Drive to a restaurant and order a meal. Go into the kitchen and prepare food. Phone for pizza. And so on.) Your mesolimbic dopamine system will establish a connection between: the sequence of behaviors you performed just before your hunger was satisfied; the internal state of hunger; and the behavioral outcome (actually having had your hunger satisfied). It has constructed an ARN.

This dopaminergic system mediates the acquisition—the learning—of behaviors that result in our getting food when we're hungry, water when we're thirsty, and sex when we're lusty. It mediates homeostatic functions—like changing our clothes when we're too warm or too cold—as well as our ability to accomplish other life-enhancing tasks.

Beyond this, activation of the mesolimbic dopamine system enhances life: It produces the *feelings* of pleasure that compel us to move toward those things in our physical and social world that have adaptive value. It is the "positive outcome" component of an ARN.

Another dopamine system—the *nigrostriatal pathway*—flags stimuli in the environment that are reliable predictors of rewards. For example, they ensure that when we see the Golden Arches, we know that we have found food there before when we were hungry. Other nigrostriatal cells track our progress as we zone in on the reward.

A third dopamine pathway—the *mesocortical pathway*—projects to the prefrontal cortex, the leading edge of your brain, just behind your forehead. This is an area where planning—mentally modeling future behavior—takes place. The mesocortical dopamine system facilitates our ability to pay attention to salient environmental features—things that are novel as well as those that are associated with getting the goods in the past. Take the hunger scenario we mentioned earlier. If there are Golden Arches in the environment (previously flagged by the nigrostriatal cells)—and we're hungry—mesocortical dopamine neurons direct our attention toward them.

## CALIBRATING YOUR LEVEL OF ALERTNESS

A third monoaminergic neurotransmitter, noradrenaline (or norepinephrine), mediates your level of alertness. You're probably familiar with the idea of an adrenaline rush: the sudden flooding of your system with adrenaline—a monoamine "juice" secreted by your adrenal gland—that makes your heart race, your breathing quicken, and your metabolism speed up in an emergency. You're hiking a trail and hear the

chilling rattle of a diamondback; you're driving along a freeway and see flashing lights; you get out of your car in a park-
ing garage late at night and hear footsteps approaching.
Adrenaline primes you for fight or flight.

What you probably don't realize—because these things
happen in a blur—is that your intelligence system went
through several steps even before the adrenaline kicked in.
Your sensory systems registered some novel stimulus in the
environment (flashing red lights, rattling sound), which produced an internal state change (fear, apprehension). As your
internal state changed, noradrenaline neurons in your sympathetic nervous system sent an alert to your adrenal gland,
which then deployed the adrenaline. Your intelligence system searched for ARNs that had solved this kind of problem
in the past, and you either pulled over your car or stopped
dead in your tracks on the hiking path, heart pounding.

In other words, the arrival of the police car (adrenaline)
follows the emergency response (coordinated by noradrenaline) to the dispatcher (your adrenal gland). And the 911 call
was placed by your prefrontal cortex. What's the link between your prefrontal cortex and the midbrain—specifically
the locus ceruleus, where the noradrenaline neurons are clus-
tered? The same mesocortical dopamine pathway we discussed earlier. Lights flash; the diamondback rattles his tail.
Signals from your prefrontal cortex are routed to your locus
ceruleus, producing increased activity in your noradrenaline
neurons. At that moment, you stop whatever you were doing,
and your attention is wholly focused on the novel stimulus—
in this case, a threat. Instantly, your behavior changes.

## UPS AND DOWNS: CALIBRATING THE SYSTEM TO INCREASE VIABILITY

You can see how intimately serotonin, dopamine, and noradrenaline work together to produce global calibrations of your intelligence system. Although chemically quite similar, they have carved out different territories. And yet they function as a kind of family, a monoamine mafia, making offers that your system finds hard to refuse. Here's an example. All three of them have been implicated in cases of clinical depression. In fact, all antidepressant medications produce an increase in the level of activity in one or more of the monoamines. (And all addictive stimulant drugs, such as amphetamine, cocaine–even caffeine–act by increasing the level of activity in monoamine systems.)

What are the symptoms of a clinical depression? A depressed mood for one. A diminished interest or pleasure in most activities. Insomnia. Fatigue. Feelings of worthlessness. Impaired concentration. Sometimes feeling agitated. Sometimes feeling slow, listless. What's the explanation?

Let's analyze the information that your behavioral intelligence system has to work with. Your record of your life is made up of a series of ARNs that memorialize salient events–some good, some bad.

These ARNs are the archive your system sifts through when you find yourself in a similar situation in the future and are trying to identify the best behavioral path. But they are also part of your personal timeline, the episodes that constitute your autobiography. They record ups and down, progress and reversals. And so this outcome information provides a perfect ongoing account of the viability of your life's course.

Imagine this episodic memory record as a sequential string of positive and negative values, each weighted by the magnitude of the associated life event. (The birth or death of a child, for example, would obviously weigh more heavily than passing a driving test or having a bad day on the stock market.) The sum is either positive or negative. If it's positive, the trajectory of your life is viable. You're on course, and the behavioral energy you are expending is getting you the goods you need in the biological marketplace. But if it's negative, a red flag goes up, suggesting that the current course of your life may not be viable in the long run; you are either expending energy without achieving your goals or you have suffered a significant loss. Because behavior is so enormously expensive energetically, the best thing a person in this situation can do is to stop what he has been doing, reconfigure his life, and try to formulate a more viable trajectory into the future.

But how would your behavioral intelligence system know that it should make deep budgetary cuts in your energy expenditure? And how would it implement the change in direction? As it turns out, your intelligence system has something very much like an internal accountant-cum-financial consultant with the functional capacity to assess this ongoing record of outcome values–to do the math on your energy bills, cut your losses, and put you on a new path to solvency in the marketplace. What's more, it also has the power to slow or shut down the behavioral motivation and generation machinery to help you get off a dead-end path and get you back on your feet. We mentioned it earlier: the basal ganglia system.

It's perfectly positioned between the neocortex–which

stores the ARN record—and the motivational centers of the life history regulatory system that drive the planning and initiation of behavior. One of its primary functions is sequencing behavior—arbitrating between the various competing possibilities represented in your neocortex, and then selecting an action. It's also the core machinery for the generation of behavior. In effect, the basal ganglia is the only system that would "know" when you should abandon your current strategy, and would "know" how to change your life course.

Obviously, slamming on the bioenergetic brakes and grinding your entire machinery to a halt is an extreme measure. Even when your life is careening toward the margins of viability, your intelligence system has a series of emergency maneuvers—course corrections—that might get you back on track.

We've already described the first of these corrections—or recalibrations—of your system: the Alice effect. If you've experienced negative outcomes in the social world, a decrease in serotonergic activity shifts your social position to a level that is more likely to produce fruitful bargaining in the biological market. You become smaller, more submissive (or humble). But suppose that doesn't work. Suppose there is no positive shift in your viability. What happens then? Immediately, your arousal systems would be activated: The level of activity in your noradrenergic and dopaminergic systems would increase.

The increase in noradrenaline and dopamine activity would crank up your mental processes and increase your level of behavioral activity. For our ancestors, problems like finding food or shelter might have been solved by mentally reviewing past foraging expeditions and then exploring more territory.

Basically, you're doing an inventory, revisiting scenes from your recent past, and reassessing their impact on your life—trawling for clues that might shed light on your current predicament and mentally modeling a behavioral path out of your difficulties. This process might well lead to a solution—but it might not. Ruminating about the events of your life at a time when your life has not been going well means that—unavoidably—you're reactivating representations of negative outcomes. All the news is bad news. What your system needs is an injection of good news, but it continues to broadcast the latest disasters. And to top it all, your noradrenergic neurons are sending out alarm signals: You're in trouble. Inevitably, this recalibration of your intelligence system registers as "anxiety." It may not *feel* good, but it's not supposed to. It's designed to kick-start your problem-solving circuits and get you out of trouble. In other words, it's adaptive.

But now, the downward spiral may intensify. If this dopaminergic/noradrenergic recalibration fails to lead to a plausible solution, both systems ramp up, resulting in a further increase in the rate of mental modeling and behavioral activity. You simply cannot stop processing your problems, and your sleep cycle is invaded (in other words, you develop insomnia). That's accompanied by a further decrease in serotonergic activity, which has the effect of lowering your self-esteem. Now you are suffering from what your physician would call "agitated depression."

This recalibration of your behavioral intelligence system is immensely costly—a hemorrhaging of energetic resources. And if this higher rate of processing doesn't pull you out of the doldrums and set you on a new path—it's all wasted: another loss registered by your internal accountant. You've now reached the point where your system, in a sense, files for Chapter 11, reorganizing your remaining assets to protect you from bankruptcy.

Faced with continuing costs, and in the absence of any income, the basal ganglia apply the brakes, slowing the dopamine and noradrenaline systems to little more than an idle. Your emotions are dulled; cognitive and behavioral activity levels hit bottom. And again, your serotonergic system downsizes your sense of self-worth. You are small. You are in a clinical depression.

Amazingly, even this most extreme state of calibration of the system probably evolved because it was adaptive. So, for example, if you were an ancestral human who was being exploited by another individual or group of individuals, a complete behavioral shutdown could abruptly force a re-negotiation of the inequitable social relationship—I am at my wit's end. I need support. I am carrying too much of a burden.

In the contemporary world, a major depression seems less likely to serve the purpose of ultimately enhancing over-all viability, and yet it often serves as a wake-up call, prodding people to abandon dead-end jobs and relationships. The irony is that, especially in industrialized societies, people suffering from a major depression are labeled sick—which works against their ability to find a new, promising path. (We're certainly not advocating bypassing pharmacological treatments as an intermediate step—but viewing depression, in this sense, as a "natural" process might help remove the unwarranted stigma.)

Another condition that occurs in situations of diminished viability is mania—a state characterized by an unrealistically elevated or expansive mood, increased activity and motivational levels, racing thought processes (flights of ideas), and an inflated sense of self-worth that verges on the grandiose. Superficially, mania appears to be the polar opposite of depression, but both conditions are systematic calibrations that reflect a sense of diminished social viability. In fact, mania is often found to cyclically alternate with depression (manic-depressive illness, or, as the *DSM-IV* calls it, "Bipolar 1 disorder"). Although not as well-researched (and consequently not as well-understood) as depression, mania appears to involve increased activity in dopaminergic and noradrenergic systems, and erratic activity in serotonergic systems. The inflated sense of self-worth—making yourself larger—hints at an attempt by the behavioral intelligence system to enhance your biological marketability; the increased dopaminergic and noradrenergic activity suggests a system being rapidly cranked up and subsequently depleted.

In her book *An Unquiet Mind: A Memoir of Moods and Madness,* Kay Redfield Jamison—professor of psychiatry at the Johns Hopkins University School of Medicine, and a manic-depressive—recalls an academic garden party she attended early in her career. Although she didn't know it at the time, her system was shifting from "play" to "fast forward" as she took one of a number of roller-coaster rides into mania. From the perspective of her recalibrated sense of self, ". . . I had a fabulous, bubbly, seductive, assured time." One of the other guests—who would become her psychiatrist—saw it differently. As he told her later, she seemed frenetic and too talkative. He thought she was dressed provocatively, had on much more makeup than usual, and looked manic. "I, on the other hand, had thought I was splendid," she wrote. Later, as the mania picked up speed, she wrote, "My mind was beginning to have to scramble a bit to keep up with itself, as ideas were coming so fast that they intersected each other at every conceivable angle. There was a neuronal pileup on the highways of my brain."

The most common pharmacological treatment for a mind racing into overdrive like this continues to be lithium carbonate. At one time, it was thought that lithium had no significant effects on people not suffering from the manic condition. We now know differently: It can produce lethargy and mental confusion in "normal" individuals—some of the more vegetative symptoms of depression. (Vegetative, incidentally, means "passive, like the growth of plants; showing little mental activity." In other words, you're in the doldrums: Your system is stalled.) These symptoms result from decreased levels of noradrenaline and dopamine. The implication is that lithium dampens activity in these systems.

There are many people who are manic or depressive—or both—who just don't appear to the rest of us to fit the profile of someone who is marginally viable on the biological market. We look at someone like Virginia Woolf, for example, and see a highly accomplished and successful individual whose writing has transformed the lives of many people—particularly women. From an outsider's perspective, we might ask: What on earth could she possibly be depressed about? She had genius, fame, and wealth. And yet, on March 28, 1941, she loaded her pockets with stones and drowned herself in a river near her home. The prospect of navigating another day had become intolerable for her.

And that is precisely the point: for *her*. Not for us. Whatever we might think or observe from the outside is ultimately irrelevant. All that counts is how Woolf felt on the inside. ("I have a feeling I shall go mad," she wrote to her husband. "I hear voices and cannot concentrate on my work. I have fought against it but cannot fight any longer.") The path she chose was a function of her unique circumstances; of how her system was calibrated in its formative years (she was sexually abused by her half-brother, and her mother died when she was in her early teens); and of her subsequent life history. The same is true of you. Your database for assessing your viability at any given moment is the idiosyncratic parade of ARNs that only you could create and that only you are privy to.

One caveat: Is it possible that some people have an innate predisposition to mania or depression? Of course. But our position in this book is to explore the adaptive suppleness of your intelligence system before assuming that it is genetically hobbled. It's possible, of course, that, in some people, the operations of the basal ganglia are on more of a hair trigger than for the majority of us. Perhaps their activation thresholds are exceptionally low. Perhaps some of us are born with a higher density of a specific type of serotonin receptor (known as S2A) that seems to underpin affiliative behavior and gives us a leg up in the dominance hierarchy. Perhaps. These variations in the phenotypic expression of an adaptation are entirely to be expected. As we said earlier, not all hearts or livers are identical. In the same way, certain aspects of the intelligence system will vary.

Our ideas about the role of the basal ganglia and monoaminergic systems in the recalibration of the behavioral

intelligence system (or the symptoms associated with depression and mania) are too new to have been directly tested. But
there's strong supportive evidence in the effects of an often-
maligned (but surprisingly effective) treatment called electro-
convulsive therapy (or ECT).

## ECT: HITTING THE RESET BUTTON

The odd thing about ECT is that it works for both mania and
depression. Why? Why would an identical stimulus–a jolt of
electricity–produce the same therapeutic effect in patients
with apparently diametrically opposed conditions?

Patient A is manic, with inflated self-esteem or grandios-
ity; she doesn't need sleep, talks a mile a minute, has flights
of ideas that are exhausting to follow, shops to drop, engages
in "sexual indiscretions" (as the *DSM-IV* coyly puts it), and
makes foolish business investments. Kay Jamison provides
testimony: ". . . mania is a natural extension of the economy.
What with credit cards and bank accounts there is little beyond reach. So I bought twelve snakebite kits, with a sense
of urgency and importance. I bought precious stones, ele-
gant and unnecessary furniture, three watches within an hour
of one another. . . . During one spree in London I spent several hundred pounds on books having titles or covers that
somehow caught my fancy: books on the natural history of
the mole, twenty sundry Penguin books because I thought it
would be nice if the penguins could form a colony. . . ."

Patient B feels sad or empty most of the time, has no
interest in–and takes no pleasure in–most activities, can't
sleep, is droning and monotonic, feels fatigued, can't con-

centrate or think very clearly, and entertains thoughts of death. Here's Kay Jamison again, on the flip side: "Suicidal depression, I decided in the midst of my indescribably awful, eighteen-month bout of it, is God's way of keeping manics in their place. It works. Profound melancholia is a day-in, day-out, night-in, night-out, almost arterial level of agony. It is a pitiless, unrelenting pain that affords no window of hope, no alternative to a grim and brackish existence, and no respite from the cold undercurrents of thought and feeling that dominate the horribly restless nights of despair."

These are, obviously, extraordinarily different conditions to those who experience them. What could be the link? How could you take two people—one virtually comatose, the other wound tighter than a watchspring—subject them to the same treatment, and have two "normal" individuals emerge at the end?

Any explanation would have to account for the astonishing behavioral changes, plus these facts: The side effects of the treatment are short-term memory loss and confusion, and the therapeutic effects are temporary.

We suggested earlier that depression and mania could be interpreted as dramatic—but systematic—recalibrations of your intelligence system instigated by the basal ganglia. After "reading" the latest (in this case, repeatedly negative) episodes of your autobiography (the behavioral outcomes in your sequence of ARNs), the basal ganglia step in to alter your course before the situation gets any worse. Think of the basal ganglia as your personal version of Alan Greenspan and his colleagues on the Federal Reserve Board (admittedly, a stretch). They are poring over your latest returns and financial projections, assessing the strength of your internal economy. And the numbers aren't very reassuring: Two quarters of negative growth is the definition of a personal recession. Your viability is in question. The basal Greenspan feels that action is called for. And so your system cranks up if the auguries of success in your marketplace are perceived to be good (mania) or shuts down if the indicators are negative (depression).

But suppose that the transmission of negative reports reaching the basal ganglia is interrupted. Suppose that the lines of neural communication are broken or the signal scrambled (just as a solar flare can disrupt communications networks). This is essentially what happens in serial treatments with ECT. The transmission of outcome information from episodic memory tracts is disrupted. The very information that was leading the basal ganglia to take emergency action—your Fed's entire dossier of economic indicators—has been scrambled. The memory information that was fueling and sustaining the mania or the depression is no longer available. In that sense, no news is good news. The so-called side effects of the treatment—memory loss and confusion— actually *are* the treatment. But given the model we've described, you can see immediately why the therapeutic effects might be short-lived (and, as any psychiatrist will tell you, they are).

As soon as the activity in episodic memory records has returned to pre-treatment levels—as soon as the pages in your autobiography are restored to their proper sequence—the basal ganglia will once again "read" reports of a system that is headed for neuroeconomic ruin and needs to be recalibrated. At that point, after the memory deficits fade and the mental confusion lifts, symptoms of mania or depression may recur–prompting, once again, the Caterpillar's question: What size do you want to be?

We answered that before: the right size to get the goods you need. But, on reflection, that was an incomplete answer. Perhaps we should have said: the right size to get the goods you need *at any given time of your life*. Timing is everything– as we explain in the next chapter.

# 6

## A MIND FOR ALL SEASONS

Here are some images to conjure with and set your adaptive representational networks reverberating: tumbling bacteria, bees on nectar missions, and surly grizzlies. Chances are you have been transported back to Chapter 1. There, you remember, we set out the fundamental principles that explain the design of your mind. We said that the garden is a biological marketplace, a landscape where deals are being struck—where the bottom line is paying your energy bills, and the ultimate prize is passing on genes. Solving this basic problem of thermodynamics explains the glory of the garden, and—eventually—the human mind. ("Energy," as the poet William Blake famously wrote, "is eternal delight.")

But now let's imagine that we're sitting in an almost identical imaginary garden in Australia—perhaps somewhere in Queensland. The birds and bees are still plying their trade, but there is also a local mammalian celebrity darting about: a mouse once mentioned in the *Guinness Book of Records*. It's not Mickey or Minnie Mouse—although it is rather cute and tiny, with a long pointy nose and an audacious habit of building nests in television sets and lounge chairs. Its name is Marsupial Mouse: more formally, *Antechinus*. (A marsupium, incidentally, is a fold of skin, forming a pouch in which the newborns hitch a ride and nurse on the mother's teats—as in a kangaroo.) What accounts for its appearance in the record books? *Antechinus* is the most extreme example of an intense period of reproduction and death in male mammals.

Here's what happens. As the breeding season approaches—a tightly defined window of just a couple of weeks in August—male marsupial mice, rather like athletes in train-ing, peak for the big event. Statistically, the odds of their sur-viving the next twelve months and returning for another attempt are negligible. (In nature, most mice are toast after about three or four months—owls and cats have to make a living.) So, in evolutionary terms, this is their best—probably their only—shot at genetic immortality. And it becomes a sexual Olympics: There is intense, grueling competition for access to females. The pace is unrelenting: A single copula-tion can last for twelve hours. Imagine.

At the end of it all, the male *Antechinus* is spent. This is hardly surprising. He hasn't eaten, levels of stress hormones (corticosteroids) have shot up, male reproductive hormones (androgens—mainly testosterone) are relentlessly coursing through his system at extraordinary concentrations, and his prostate has ballooned to fifty times its normal size. As a con-sequence of all these biochemical insults, his immune system collapses. Within a week or so—by the end of August—he is dead. Sometime in the first two weeks of September, his off-spring are born.

If mice were men and this were a movie, it would seem

the blackest of stories—a *film noir,* in which no child ever meets his or her father. And yet there is a reasonable explanation: Underpinning the plot is solid evolutionary logic. It's known as life history theory, and it's the subject of this chapter—a chapter that brings us full circle in our story of how your mind works.

Life history theory logically flows from the laws of thermodynamics—the energy constraints—we discussed in Chapter 1. There are three basic life functions that require energy: maintenance, growth, and reproduction. And there is a fundamental, limiting principle of allocation: If you use energy for one purpose, you can't use it for another. Life history theory is about trade-offs. You can see how this works with *Antechinus.* After a certain point—the beginning of those two frenetic weeks in August when *Antechinus* puts all its chips on reproduction and rolls the dice—maintenance and growth are abandoned.

Something similar happens with Pacific salmon. They leave their birth streams for the open ocean, develop and grow for two or three years, then return to their natal breeding grounds. You have probably seen the wrenching videos of salmon, desperately thrashing upstream to spawn. They don't eat from the moment they enter their home river; instead, they reabsorb their own bodily tissues and break them down to make eggs, fueling the reproductive imperative. As with *Antechinus,* the salmon's hormonal systems shift into overdrive. Adrenal and pituitary glands grow enormously and corticosteroid production amps up, shutting down other vital metabolic activities and fatally compromising the salmon's immune system.

Wherever you look in the natural world, you find examples of this all-or-nothing, big-bang breeding. Mayflies that dance briefly at their gossamer, aerial weddings, then fall to the water below. Bamboos that save themselves for 120 years before suddenly flowering and dying. The annuals in your garden that bloom, then die.

This kamikaze behavior is known as semelparity, which basically means reproducing in one grand, glorious—and suicidal—moment. It happens when the odds against a successful encore are overwhelming. In a sense, the actuary of natural selection has assessed the prospects of *Antechinus'* survival for another year, or of the salmon's capacity to fight the rapids again, and says: I don't think so. Seize the day.

In contrast, the kind of life history with which we are more familiar is called iteroparity. (Think of iteration—repeating a behavior—and the Latin verb *parere:* to bear offspring.) Iteroparous creatures—like rats and cats and monkeys and humans—can reproduce repeatedly, tend to have fewer offspring during each birth season (compared to the thousands of eggs that salmon lay), invest more of their own ongoing resources in the survival of their young, and have a relatively long post-reproductive life.

These radically different journeys are under the control of the life history regulatory system (LHRS). In humans (and other animals), the LHRS is primarily composed of the hypothalamus, the endocrine glands (which are regulated by the hypothalamus), and an area of the midbrain called the median eminence. What does the LHRS do? It supervises and regulates the scheduling—the unfolding—of our lives according to information contained in our genes (which are technically part of the LHRS).

If you've been following the reports of the Human

Genome Project, you probably know that the genome contains the entire set of hereditary instructions (DNA) for building and running you—and every other individual in every other species. Every human has a genome—broadly similar but different in the details (with the exception of identical twins). Every mouse has a genome—which differs from the genome of every other mouse. Every honeybee has a different genome; every salmon, every *Antechinus*.

Although this is not the place to launch into Molecular Genetics 101, it might help to have some sense of the basic machinery. Each cell in your body (with the exception of red blood cells) has a nucleus. Packed into the nucleus is a "library" of chromosomes, with genes strung out along their length. Different species have different numbers of chromosomes. Humans have forty-six (in twenty-three pairs); dogs have seventy-eight; cats have thirty-eight. Some of the genes (in humans, a surprisingly small proportion, it now turns out) "code" for—in other words, direct the manufacture of—proteins from the twenty amino-acid building blocks. Almost everything in your body or brain—your skin, hair, muscles, enzymes, hormones—is protein (or derived from protein).

The genetic code that specifies proteins is elegantly simple (at least in hindsight). In the DNA molecule—the famous double helix, rather like a ladder twisted into a spiral—the "rungs" linking the two sides of the ladder are made of pairs of chemicals called bases. Adenine (A) always links to thymine (T), and cytosine (C) links to guanine (G). If you ran up the ladder looking only to one side, you might see a sequence of bases like AGTCCGCGAATAC. (Go back and do it again, looking to the other side and you would see

TCAGGCGCTTATG.) These base "letters" constitute a genetic alphabet–a code. A triplet of letters–like TGG, for example–tells your cellular machinery to reel in the amino acid tryptophan and start building a sequence with it. As another triplet rolls by on the genetic assembly line, another amino acid is snagged and butted up to tryptophan. Pretty soon, stringing amino acids together like beads on a necklace, you have a protein. And then you're ready to tackle the really big construction projects like hearts and lungs and skeletons and eyeballs–and brains.

All this construction work has to be scheduled. There's no point having the finish carpenters or the tilers in the house before the footings have been poured or the framing has been done–just as there's no point entering menopause before you've even reached puberty. If you're building a house, the solution is simple: You hire a general contractor who makes sure that the various specialist subcontractors turn up on time, in sequence, and do the job within the budget. In a sense, the LHRS is your general contractor. And it's part of your genetic package.

That package is defined the moment your father's sperm fertilizes your mother's ovum to create a cell called a zygote. The zygote divides and grows and–after about a week–is implanted in your mother's uterine wall as an embryo. Nine months later, you are born. A year later you take your first shaky steps. Perhaps twelve years later, you have your first period. And then your first child. Somewhere in your fifth decade, your prospects of childbearing radically diminish. There are similar milestones for males. All this is part of your developmental schedule.

You may be under the impression that the genome

tightly defines the limits of who we are and what we might become, as if a carpet of behavior is being inevitably unrolled, gene after gene. The genome, after all, has been variously described as "the book of life," our "genetic blueprint," the "map" of human possibilities. The implication is that our future is immutably inscribed here: that the sentences of the book are to be read; the lines of the blueprint are to be traced; the contours of the map are to be followed—without deviation. That life, in other words, is a strictly constrained verbatim performance of the genome.

It's true that certain genetic instructions—like the mutation on chromosome 4 that causes Huntington's chorea—may lead inexorably to diseases. In the case of Huntington's, the book of life has a devastating misprint. One "word" in which the coded message of DNA is written—CAG, CAG, CAG . . . the triplet that codes for glutamine—begins to repeat, to perseverate. It's as if the famous opening lines of *Moby-Dick* had been: "Call me Ishmael ishmaelishmaelishmaelishmaelishmael . . ." and on and on for forty or so repeats. Inevitably, this kind of biochemical stammering in the genome has behavioral consequences.

In the same way, certain physical traits—like eye color or Sophie's cute nose ("*So* like her mother")—are attributable to the handing down of a genetic recipe for the production of blue eyes or cute noses. In the genetic swap meet of sex, the instructions for making a child come from her mother's and father's gametes (the sex cells), which, in turn, retain elements of the genetic legacy of *their* parents, and their parents' parents, and so on. That much, we understand.

But when Sophie—blue eyes, cute nose and all—begins

to behave in the biological marketplace, life gets much more complicated. Analogies like "blueprint" or "map" don't work very well as a description of how individuality unfolds. Perhaps music is a better example.

The classical music of all cultures—from Western symphonies to Indian ragas—has a quite formal structure. And if you've ever practiced scales, or labored through a rendition of "Chopsticks," you know that music has many of the elements of a language. The notes are like letters. String several notes together, and you have a phrase or a chord (a musical word). String together the phrases and chords, and you have a passage or a movement (a sentence). Eventually, you can create a sonata or a symphony (a story or a book) from these basic building blocks. Now think of jazz—perhaps an improvisational flight by Charlie Parker, John Coltrane, or Miles Davis, for example. The specific sequence of notes they played may never have been heard before. Underlying the music was a rigorous use of the musical alphabet—they spoke the language, could write the sentences, understood the grammar—but they used it in a uniquely different way.

Your LHRS is underpinned by a formal (evolved) notation, written in your genes. The stages of life unfold, rather like a symphony. But, at any moment, environmental inputs can affect how—and when—you play a certain passage. At any moment, you have the capacity to take your instrument and improvise a sequence (write a story) that only you could have composed. Maybe, as you face the music of the reproductive stage of life, you decide to study for a graduate degree instead of starting a family. Or maybe, for you, the goods consist of celibate life in a seminary. Maybe, as a childless, post-reproductive female, you decide to adopt a child from a different culture. The point is, you have choices. You can improvise a solo. Life is jazz.

In life history theory, there are two ways of doing business—ballistic and conditional. What does ballistic mean in this context? Suppose you have a cannon and light the fuse. The gunpowder produces an explosion, which launches the ball. The ball follows an arc, then falls to earth. Once you light the fuse, there's nothing you can do to control the cannonball's trajectory. Everything that you did prior to lighting the fuse (like the path you took to get your cannon in place, or measuring the amount of powder used, or setting the elevation of the barrel) may have been conditional.

In organisms, lighting the fuse is equivalent to "firing" up hormones, which then launch ballistic behavior. Salmon battle upstream to birth their offspring. Mayflies dance the wedding waltz. *Antechinus* competes in the mouse Olympics. There is nothing to be done, at this point, to change the trajectory of the hormonally induced behavior. Pacific salmon can't turn tail and head for the open seas again. *Antechinus* can't retire from competition. And mayflies can't sit out the dance.

Ballistic processes are initiated by pacemakers within cells; they operate independently of external environmental signals. Many of the life history jobs performed during a life stage, such as the growth of bones during the developmental period, are mediated by ballistic processes. Conditional processes, by contrast, are triggered by cues from the environment, and virtually all major life-stage changes are predicated upon the appropriate environmental input. For example, an environmental cue triggers the salmon's LHRS "decision" to transition to the reproductive life stage, and start swimming up the river where it was born, to spawn. But once that trigger has been pulled and the powerful hormonal machinery of the LHRS has kicked in, the salmon—literally—goes ballistic. The mayfly, too, goes through several developmental stages before the final, desperately brief ballistic episode (but note that even the launching of this terminal phase depends on environmental signals—changes in light and temperature).

Although the more dramatic examples of ballistic behavior occur in semelparous organisms, all life histories—whether semelparous or iteroparous—shuttle back and forth between conditional and ballistic episodes. That makes obvious sense for iteroparous creatures like ourselves who may transition several times from mating mode to reproductive mode to parenting mode. The transitions may be environmentally sensitive (conditional), but once triggered, they are largely ballistic: Zygotes divide and become embryos, which implant and continue through a programmed series of developmental stages.

By now, you can see how crucially important the LHRS is to the unfolding of your life. But where does the mind come in? After all, most life forms get through life without a mind. Plants and trees in the garden have essentially a no-frills life history regulatory system that sequences their existence from seeds to senescence. Hormones, responding to environmental cues—like ambient temperature and hours of daylight—orchestrate their schedule. If it's spring, they flower. If it's fall, they shed leaves. If it's unseasonably hot and energetic resources are at a premium, they can reduce their expenditure—again by shedding leaves. But that's about it. The plant LHRS *is* the intelligence system. Roses and red-woods have no apparatus for assessing their situation, comparing it to previous emergencies, finding a historical match, and executing an appropriate response. They don't have a menu of possible actions–like "shuffle inside the house" or "jump into the pool"–because they don't have a behavioral intelligence system.

If you lived in the cartoon world of Disney, where trees *do* pick themselves up by the roots and shuffle inside the house, this would all change. A no-frills model LHRS wouldn't be enough. You'd need to evolve a plastic system that allowed you to interpret information, make predictions, make behavioral choices, and weigh the energetic costs and benefits.

A behavioral intelligence system is a complex of evolved adaptations of the life history regulatory machinery: It makes the LHRS a better mousetrap. And mixing metaphors, the icing on the cake is the cortex, with its extraordinary ability to hold in representations the story of your life (from your unique perspective).

You'll remember from earlier chapters how well-connected the hypothalamus (and hence the LHRS) is to the decision-making circuitries of the basal ganglia. Positioning yourself in the biological marketplace, becoming larger or smaller, keeping your energy bank balance in the black–all these intertwined tasks stem fundamentally from the mandates of the LHRS. It sets the agenda. For animals (in contrast to the life history of plants), that agenda is influenced by social factors–by market demands and opportunities. Here's a telling example.

Cichlid fish live in shallow pools and estuaries of Lake Tanganyika (East Africa). At any given time, about a third (a range of twenty to forty percent) of the males maintain territories. Territorial (dominant) males are brightly colored and have mature testes: Reproductively, they're ready, willing, and able. Subordinate males–those that don't have territories–are a different kettle of fish. They tend to swim with the females, are drab in color, and are reproductively inactive.

Now imagine that the rules of the game are changed–either by scientists manipulating the environment or by natural circumstances. Sometimes, for example, predatory birds selectively pick off the more visibly florid territorial males. Sometimes, a hippopotamus wallows through the cichlid world, chaotically disrupting the established territories. Then what happens? It turns out that this fluctuation in the habitat wreaks havoc on cichlid social status, which, in turn, regulates growth rate. Suppose a dominant male cichlid is snapped up by a bird. The subordinates then contest the vacant territory. In the space of about a week, the winner climbs the social-status ladder, cranks up its growth rate, and becomes reproductively active. Even the hormone-releasing neurons in the victor's brain cells increase in size. (Vice versa: When a cichlid switches from being territorial to nonterritorial as a consequence of a change in social status, the cells shrink.) All this activity is regulated by the hypothalamic-pituitary-gonadal axis (a kind of hormonal powerhouse), which is their LHRS. You can see clearly the trade-offs we have described. Cichlids that are territorial (dominant) invest in reproduction. The subordinate males bide their time, investing in maintenance and—when their chance comes—growth. Then they shift into reproductive mode.

There is an even more exotic version of this social regulation of life history trajectories—also in fish, called wrasse. Their mating system is harem-based polygyny—one male mates with many females. But only a few of the largest males—the bluehead wrasse—have harems, leaving few or no females for the smaller subordinate males—which are, perhaps appropriately, colored yellow. As with the subordinate cichlids, subordinate wrasse seem doomed to a life of being ineffectual understudies, waiting for the great man to retire. Which could be a long time coming. What to do? Well—reproductively speaking—a small male might be better off as a female. At least then there would be a chance of leaving a mark in the wrasse genealogies. As luck (or natural selection) would have it, the wrasses' LHRS can accommodate that change. Small males would rather switch than fight: They morph into females. But that's not the end of the gender-bending dance. If a dominant male wrasse dies, top billing is up for grabs. The small male-cum-female reverts back to its maleness (including a quick costume change from yellow to blue) and takes over the territory—and the spoils. One final twist. Surgical removal of the gonads (the sex organs) *doesn't change* this story. In other words, becoming a dominant male wrasse doesn't require—you should excuse the expression—balls. It is a function of change in the social environment.

These examples of social modulation of reproductive physiology are not just fishy stories; they occur in primates, too. Take the mandrill, for example. Although it looks like a baboon (they're closely related), it's actually the biggest monkey (part of the same family as rhesus macaques and colobus monkeys). You may have seen one in a zoo. It's the most colorful mammal, with a long blue snout and cheeks, a bright red nose, and pale yellow beard and mane. Dominant mandrills are much bigger than females, maintain harems, and are immensely powerful animals with huge canine teeth. Growing up as a juvenile—or living as a subordinate—male mandrill in the presence of such a formidable paterfamilias is daunting. How daunting? Enough for the LHRS of a young male mandrill to delay its sexual development until there is a sporting chance of striking out on its own.

Obviously, we're not cichlids, or coral reef fish (wrasse), or even mandrills. But we do have a shared evolutionary history. And natural selection is fundamentally a frugal (one might almost say cheap) operation: If it can make do with a quick fix of a system that it's already spent millennia polishing, it's unlikely to invest in some new, speculative enterprise without good reason. So: Given that social factors in all these examples regulate the physiology, psychology, and behavior of an individual, it stands to reason that human reproductive physiology is likely to be socially modulated. How? Well, suppose we told you that the regular presence or absence of a father figure in the home apparently shifts the age of menarche (first period) and first sexual activity in daughters? If a girl's father is absent, she may reach menarche and become sexually active at an earlier age—as many as two years earlier—than she would if he were present. Why? Economics.

All of us have evolved intelligence systems that constantly have one eye on the bottom line. For a female child whose father has abandoned her (or, perhaps, been killed in some ancestral dispute or hunting accident), the terms in the equation of her survival may be radically altered. At one stroke, her income might fall to a fraction of what it once was. Her mother and other relatives may try to make up the shortfall. But the fact remains that she could now be living on the margin. Even in contemporary societies—with social services and child support—her future could be tenuous. What can she do?

Let's step back for a moment. If you're a girl whose father is present, the odds are that you'll continue receiving his parental investment at least until you reach sexual maturity. (There's a good reason for this: In the fundamental marketplace currency of genes, his investment doesn't produce dividends until you give birth to his grandchild, ferrying his genetic legacy into the future.) From the perspective of your behavioral intelligence system, all the market indicators so far have been positive. When you were hungry, your parents supplied food. (In ancestral societies, perhaps your father went on a hunt to supplement your mother's gathering; in contemporary societies, maybe he racked up overtime or moonlighted on a second job.) Every bioenergetic check you have cashed has been covered by the parental bank. On your personal Dow Jones, your stock is riding high. Because your parents have underwritten your prolonged adolescence—endowing you with a longer childhood during which you accrued resources (strength, size, skills)—they have given you a head start in the mating market you are about to enter.

Now suppose you are a girl on the cusp of transitioning from prepubescence—but your father is absent. In ancestral tribes (and contemporary tribal societies such as the !Kung or Ache), your position is probably more tenuous. There is a real possibility that you might not survive. How does your LHRS respond? By shifting you, hormones and all, to the reproductive stage so that you can attract the provisioning and protection of a mate. The environmental cues associated with father absence apparently trigger an unusually early surge in gonadal and adrenal hormones, leading to early menarche.

Notice that we said "probably," and "seem to suggest," and "apparently triggers." There's a reason for this caution. Throughout this book, we've taken some pains to remind you that you are a unique individual, dealing with unique circumstances and responding in a unique fashion. Different behavioral intelligence systems—although grounded in the same evolutionary logic—produce different solutions to life's challenges. Is it possible that ancestral—and contemporary—situations in which a single mother and her extended family supply her child's energetic needs might mitigate the physiological shift to earlier menarche? Of course. There may be individual cases where father absence doesn't constitute an energetic emergency. But, in general, a downturn in your energy portfolio will trigger a search for new investors: You enter the mating market.

Virtually all major life history changes, such as these transitions from the developmental to the reproductive stage of life, are conditional: They depend on environmental inputs. In iteroparous organisms like us, there are no hormonal border guards issuing one-way visas. We can shift, for example, from reproductive mode to parenting mode and back again. The mother of a two-year-old does not suddenly cease to be a sexual being. At any moment, her social world may cue a shift from parenting to mating mode or vice versa. The cue can activate ARNs that recall her sexual activity, or her experience as a first-time mother. She can recall moments when she was sexually transported, moments when she first nursed. And each remembrance is colored by her physiological response. Her social environment can produce conditional changes even *within* a life stage.

Amy Irving, former wife of the film director Steven Spielberg, tells a story about her experiences as a young mother that illustrates this point well. Irving was bathing their son, Max. Because he was hungry and wanted to nurse, Max started to cry. Spielberg was filming *The Color Purple* at the time and, for a particular scene, he needed the sound of a child in distress. As the story goes, he quickly set up sound equipment in the bathroom and recorded Max.

Fast-forward to more than a year later. The Spielbergs attended the premiere of the movie. Although Irving had since weaned Max and had stopped producing milk, the moment she heard her son cry on film, she said, the milk began to flow. Why? By now you can figure out the answer easily. Think back to earlier chapters, and step into Irving's shoes. There is an environmental input (the sound of Max crying). This constitutes a problem to be solved. Your mind accesses adaptive representational networks that offer potential solutions. (When Max cried before, did you: a) take him out of his crib and rock him; b) feed him; c) hand him off to Steven; d) call your mother; e) follow some antiquated child-rearing text that advocated turning a cold ear to the sounds of distress; f) summon a nurse?) What worked best? Apparently, in this case, b) was the key. So when Amy Irving heard Max cry, she was conditioned to respond by lactating. And the milk began to flow.

Environmentally induced LHRS effects like these occur regularly throughout our lives—and not just for women. Social factors can influence physiological changes in men as well (just as they did in cichlid fish and mandrills). High-status males, in general, have higher levels of testosterone. And usually, the high status is a consequence of having been the winner in some kind of social contest. (Performing in the biological marketplace seems to be an important key. A criminal trial lawyer, for example, will often have a higher level of testosterone than, say, a tax or patent attorney.)

Sample the testosterone levels of men before and after a sporting competition, and you find that the winner's testosterone levels have risen, while the loser's levels have fallen. (There are obvious parallels here with serotonin.) The intelligence system of a player registers the change in the individual's position within the biological market as a result of the outcome of the competition. Is he dominant or subordinate? Should he be bigger or smaller on the Alice scale? And what are the behavioral consequences? If this had been an ancestral tribal war, for example, the hormonal surge of testosterone in the victor would have prepared him for reproduction. On the other hand, the LHRS of the vanquished would counsel caution: live to fight another day.

So powerful are these environmental cues that they can affect even the *audiences* of social confrontations. For example, male spectators at a soccer game reportedly experience a testosterone lift when their team wins. Although this hardly qualifies as a controlled experiment (you can imagine all sorts of potentially confounding variables—like alcohol intake, for example), careful studies of embattled cichlids suggest some intriguing parallels. The basic idea that the researchers set out to explore is simple and mirrors the theme of this chapter: Social interactions influence an individual's LHRS and the production of hormones. But what happens to hormone levels in bystanders, rather than in the combatants? What happens to your testosterone levels if you're a male audience member watching a World Wrestling Federation bout?

In the cichlid version of the WWF, two fish were put in solitary to minimize the influence of prior social experiences on testosterone levels. (They were separated by an opaque partition, and cruised their underwater dressing rooms for a week.) Bystander cichlids watched the pair through one-way mirrors for several days. Then the partition was raised; for an hour, the fish struggled for dominance. Testosterone levels in the bystander cichlids increased. Why? Essentially, it seems they were getting ready to rumble. The hormone flux sharpened their social skills and readied them to challenge the combatants—as Shakespeare puts it in *Henry V:* "Stiffen the sinews, summon up the blood, disguise fair nature with hard-favour'd rage; Then lend the eye a terrible aspect."

LHRS hormones like testosterone orchestrate changes in physiology, psychology, and behavior at each stage of your life. Shakespeare, again, seems to anticipate this when he gives the character Jaques (in *As You Like It*) one of his greatest speeches—now called the Seven Ages of Man. "All the world's a stage," he begins, "And all the men and women merely players; They have their exits and their entrances; And one man in his time plays many parts, His acts being seven ages." And then he lists the transitions—the passages—from the infant to the "whining school-boy"; from the lover "sighing like furnace" to the soldier "sudden and quick in quarrel"; from the justice "in fair round belly" to the aging pantaloon "his big manly voice, Turning again toward childish treble"; and, finally, to the last scene of all–"second childishness, and mere oblivion, Sans teeth, sans eyes, sans taste, sans everything."

In essence, this is a chronicle of shifts in LHRS-mediated hormonal levels triggering new stages in the life span: weaning, puberty, mating, parenting, and so on. And often the same hormone is responsible for the many diverse characteristics of each life stage. This hugely complex parallel-processing job is possible because the release of a single LHRS hormone into the general circulation produces multiple distinctive effects, simultaneously activating different molecular receptors at different locations throughout the body and brain.

Think of some of the myriad effects of testosterone on the physiology, psychology, and behavior of a pubescent human male. First, there's an increase in the size of the penis and testes, a decrease in subcutaneous fat and an increase in muscle mass. (The genesis of a lean, mean fighting machine.) Testosterone also produces a thickening of the vocal cords (after the often-embarrassing in-between stage when a voice breaks and deepens to adult registers) and a sprouting of body hair. These physical effects are the result of the hormone acting at countless receptor sites in different target tissues throughout the body.

This ability of one hormone to orchestrate different effects is called *pleiotropy* (from the Greek words meaning, roughly, multiple changes). As life progresses beyond the reproductive years, this adaptive characteristic can become "maladaptive"–in which case it's called *antagonistic pleiotropy.*

For example, a high level of testosterone in early male adulthood is adaptive; but as time goes on, high testosterone levels may lead to prostate cancer. As we said at the beginning of this chapter, life history theory is fundamentally about the economics of survival. And hormones are the agents that mediate the trade-offs that an organism's LHRS and behavioral intelligence system put money on at any given moment. A testosterone surge in an adolescent that boosts the drive to reproduce and the leveraging of genes into the future stimulates a chain of events that can lead inexorably to an earlier death. It's hardly surprising that the French call sexual climax *le petit mort*. You might say that death is the price we pay for sex.

The LHRS is a smart gambler. It brings to the table a long (evolutionary) record of success. And it "knows" how to hedge its bets if circumstances change. Like the card-playing gambler in the Kenny Rogers song, your LHRS knows when to hold 'em and when to fold 'em. It knows when to walk away and when to run. If you're a semelparous organism such as *Antechinus* or the Pacific salmon, you bet the ranch; if you are an iteroparous organism, you're content with a series of smaller winning hands over time. If you are bioenergetically flush, your LHRS will hold 'em and bet on your survival. But if the environmental deck is stacked against you, your LHRS will fold and wait until you are dealt a better hand (like the female grizzly delaying a shift to the reproductive stage when her fat reserves are too low).

Think about some of the adjustments that have to be made within the reproductive life stage. You have a finite amount of energy. How do you spend it? How much goes to parenting effort—raising children to a viable age, at which they can reproduce and pass on some of your genetic profile? And how much do you allocate to renewed sexual activity and the potential creation of a new offspring?

Each of these options is associated with very different psychological and behavioral tendencies, which are brought about by very different hormonal profiles. And even within these distinctive stages of your life history, there is further fine-tuning by the LHRS. For example, it would be a waste of energy for the LHRS to induce high levels of sexual desire and behavior in a woman who was in the nonfertile part of her menstrual cycle. The system evolved so that the same hormonal impulses that induced the critical physiological event–ovulation–also induced the psychological state of sexual desire.

We're not suggesting, of course, that human female sexuality is strictly limited to times of maximum fertility. But theory predicts–and the evidence confirms–that this should be the part of the monthly cycle when women are most receptive. And that receptivity should show up in their overt behavior.

Suppose you wanted to design an experiment to test whether the way young women dressed correlated with their behavioral choices–like having sex. How would you go about it? Well, you could select a couple of very different activities–like going to a dance club or attending a lecture. Common sense predicts that the women attending the lecture would be more conservatively dressed than those going dancing. And you could test this by taking photographs, computer-digitizing them, and measuring the area of bare skin (front and back) on display, recorded as a percentage of total body area. What you'd find (and researchers have done this) is that, as expected, the women going dancing reveal significantly more bare skin than those attending the lecture.

Of course, you could make all sorts of ad hoc speculations about why this might happen. You could suggest, for example, that the women at the dance club are using their physical atributes to attract mates. You could just as easily suggest that the women at the lecture are also active in the biological marketplace, but are using conservative dress (and attendance at a more cerebral activity) as a filter to screen for potential partners. Or maybe it's just more conducive to dancing if your clothes are unrestrictive, and maybe it would be uncomfortable (even embarrassing) to attend a lecture in a little black number by Dolce & Gabbana. Maybe the reason for going to the dance/lecture was simply to move/learn, with no sexual agenda. The possibilities are endless—and the speculation fruitless. If you wanted to know if *fertility* had anything to do with behavior and dress, you'd have to go a step further.

What would you do? Refine the experiment in two crucial ways. First, you would take saliva samples from the women entering the dance club and the lecture hall. And second, you would have a control group (also supplying saliva) of women at the dance club and the lecture who are using oral contraceptives. Why the saliva test? Because saliva can be tested for levels of estradiol (an estrogen, or hormone, that is a prime mover of life-stage changes in women—like testosterone in men). Estradiol levels are an index of the ovulatory phase of a woman's menstrual cycle. (The higher they are, the closer the woman is to ovulation.) Why the control group? Oral contraceptives effectively dull or suppress the normal swell of steroid hormones like estradiol.

So now you have an average estradiol level for women who don't cycle in both the dance-hall group and the lecture-hall group. And you can compare the "showing skin" effect of estradiol levels within groups *and* between groups. The result? The more skin the women show, the higher their estradiol levels. The closer they are to ovulation (and a heightened potential for pregnancy), the sexier their dress. The implication is clear: Estradiol coordinates a complex of psychological and behavioral changes that complement ovulation.

What's more, this effect represents the pleiotropic orchestration of just *one* hormone dedicated to accomplishing LHRS goals. You have literally hundreds of other hormones subtly coordinating your passage through your life span.

## MANIPULATING THE LIFE HISTORY REGULATORY SYSTEM

Understanding how the LHRS works offers the prospect of manipulating the system to enhance survival. For example, various medical conditions respond to increasing or decreasing the level of a hormone in an effort to reestablish normal balance. Type 1 diabetes is a good example. It's a relatively common condition caused by the failure of the pancreas to produce sufficient amounts of insulin. In pregnant women (as we explained in Chapter 2) diabetes can also occur when the developing LHRS of the fetus is pumping too much human placental lactogen into its mother's bloodstream. Left untreated, diabetes can be life-threatening. But insulin replacement therapy makes it possible for diabetics to regain control of their systems.

The trouble is that artificially altering hormone levels can be tricky. Remember antagonistic pleiotropy? A hormone acting therapeutically in one instance might have unpredictable and harmful side effects in others. Suppose you go for your annual medical checkup, and the labwork shows that you have high cholesterol levels (or, more accurately, you have abnormally high levels of "bad" low-density lipoprotein cholesterol and low levels of "good" high-density lipoprotein cholesterol). Your physician may prescribe a cholesterol-lowering medication (probably in addition to increased exercise and changes in your diet). The next time you are tested, you find that your overall cholesterol level has dropped significantly. Good news? Well—maybe.

Here's the problem. A less formal way of defining pleiotropy is that you can never do just one thing. (Ecologists understand this only too well. What seems to be a discrete and limited change to an ecosystem—like treating crops with insecticide—can have enormous collateral effects down the food chain.) Your body is an immensely complex ecosystem. Changing the levels of just one hormone such as cholesterol (or tinkering with the availability of a neurotransmitter such as serotonin) can skew the activity of your LHRS and behavioral intelligence system.

Cholesterol, for example, plays an essential role in mobilizing energy reserves. It is a critical factor in the metabolism of carbohydrates, as well as the precursor for glucocorticoids, hormones that promote the synthesis of glucose. These hormones also mobilize fatty acids, which can be converted to glucose. In other words, if glucose levels are low, cholesterol becomes the pinch-hitter that can save the energy game.

But what happens if you have been on a medical regimen to lower your cholesterol levels? Remember, in a system this complex, you can't do just one thing. There are consequences. If you sift through enough scientific studies of the effects of cholesterol-lowering drugs and analyze the data, something startling emerges. Except for patients who already had significant cardiac risk, the mortality rate in patients who had taken cholesterol-lowering medications was actually significantly higher than that of patients who had *not* taken them. On the face of it, this seems bizarre. After all, lowering cholesterol levels is supposed to *reduce* the risk of a heart attack. And it does. And yet, for patients taking these drugs, the likelihood of dying as a result of an accident or an act of violence was significantly increased, enough to wipe out the positive effect on death rate as a result of the drug's ability to protect against a coronary attack. Why?

Think back to Chapter 1 and the grizzly bear example. If his glucose levels (a measure of his available bioenergy) have dropped into the red zone—his tank is low—his system is primed to take extreme measures to ensure his survival. And that includes the risky business of raiding your campsite. Grizzlies aside, just remember how you felt and behaved the last time your blood-sugar levels dropped when you were hungry. Were you perhaps a little cranky? Surly?

This socially induced change in disposition is part of your intelligence system's response to rapidly lowering glucose levels. If we humans (and our mammalian ancestors) evolved the capacity to become surly (and motivated to solve the problem) when we were hungry, imagine what might happen if we were on the brink of starvation.

If starvation had been a serious prospect for our Pleistocene ancestors (as it probably was—there were no fast-food outlets in sub-Saharan Africa), engaging in a risky close encounter with a dangerous animal or a competitor for food might be worth the gamble. But how would the intelligence system know that desperate behavioral measures were called for? What would serve as the signal?

Low glucose reserves are usually not life-threatening. Like other mammals, we store fat, which can be converted to glucose in an emergency. But if fat stores are low, or the capacity to tap available fat reserves is impaired, the problem becomes critical. At that point, an intelligence system should estimate the benefit of procuring food as infinitely high and the relative costs and associated risks as inconsequential. This is a high-stakes calculation: Discounting risk is as potentially lethal as starvation. So the crisis cue to the intelligence system would have to be a highly reliable index of seriously threatened bioenergetic viability—a kind of "desperation index." What would do the job? Your cholesterol level.

Like other LHRS hormones, cholesterol has widespread effects on physiological, psychological, and behavioral functions. It is the precursor (the basic material) for biosynthesis of all adrenal and gonadal steroid hormones, which, in turn, are precursors of other hormones that play essential roles in the metabolism of carbohydrates and a host of other vital physiological processes. Without cholesterol, the ability to access the bioenergy for all life processes is severely impaired. Cholesterol is also the precursor of aldosterone, a hormone that helps to maintain the electrolyte balance required for neural activity. So without cholesterol, the intelligence system's communication lines fail. As if facilitating all of these life-sustaining processes isn't enough for one molecule, cholesterol is also essential for reproductive success: The production of sex hormones—testosterone, estrogen, and progesterone—is completely dependent on the ready availability of cholesterol. Simply put: no cholesterol, no sex.

So you can see why rapidly diminishing levels of this steroid hormone would be an excellent "desperation index" for discounting behavioral risks. If that's the case, it's probably not the absolute hormonal level that is the index—rather, the *rate of decline* of the hormone. (If you recall the bacterium *E. coli* in Chapter 2, its protein circuits assess glucose ability in terms of the *rate* of molecules binding to receptors.)

## FOREVER YOUNG

During the post-reproductive stage of life, the LHRS is winding down. We call it senescence, which basically means growing old, from the Latin word for "old"—*senex*. (It's the same root that gives us the word *senate*—"a council of elders.")

The trajectories for men and women are very different. Take one example. Males can produce sperm throughout their lives. Females enter life with their entire clutch of eggs, and the number is relentlessly culled with each tick of the biological clock. A second-trimester fetus has as many as 7 million eggs; about a million remain by the time the baby girl is born, a quarter million remain at puberty, and by the time she reaches her midthirties, she has about 25,000 eggs left. At fifty—give or take a few years—the shelves of her reproductive pantry are bare. (If you do a quick back-of-the-envelope calculation, you'll see that a reproductive life phase of roughly forty years, with thirteen menstrual cycles a year, accounts for only about five hundred eggs. The rest—a thousand for each one that dallies expectantly in a woman's fallopian tubes—are presumably rejected by the body's strict quality-control procedures.)

This change in a woman's reproductive capacity—precipitous compared to the decline of testosterone levels in men—is a consequence of a drop in female production of the hormones estrogen and progesterone. Ovulation and menstruation come to an end, and a woman enters menopause. For most women, this is a bittersweet time. The good news is that you are no longer in the conception business. The bad news is that you are no longer in the conception business.

And there's the pleiotropic downside of multiple physical and psychological symptoms—fatigue, hot flashes, osteoporosis, and heart palpitations. There may be depression and, in some cases, short-term memory loss. Virtually all of these symptoms can be alleviated by taking estrogen or a combination of estrogen and progestin (a hormone similar to progesterone). But bear in mind that these are not symptoms of an illness; rather, they are symptoms of a normal progression through the life span, a transition between the reproductive and post-reproductive periods. In replacing these LHRS hormones, a woman is modifying her life history, extending aspects of the reproductive life stage and the life span.

In recent years men have also begun to use LHRS hormones to modify the length and shape of the life span—particularly with DHEA and testosterone. These hormones, like estrogen, decline in both men and women in middle age, leading to the physiological and psychological changes associated with the senescent stage of life. DHEA, which is the most abundant steroid hormone in the body, actually begins declining in men steadily from the early twenties until death, paralleling the trend in declining male sexual interest and performance.

DHEA, produced by the cortex of the adrenal glands from cholesterol, is a "mother steroid": It plays a major role in the formation of other steroid hormones, like estrogen and testosterone. In addition to being associated with a reduction in heart disease and cancer, DHEA has also been implicated as a "buffer" against brain aging, at least in men. Men with normal senility, as well as those with Alzheimer's disease, have significantly lower levels of DHEA than control subjects of the same age who have normal mental functioning.

Testosterone increases the amount of growth hormone the body naturally produces and so increases the ratio of muscle to fat. It also has been found to improve mood and to increase the sex drive in both older men and older women.

Although all of these hormones may have health-enhancing and possibly life-extending effects, they have side effects. Estrogen can cause cramping, breast pain, and weight gain, and some studies suggest that it might increase the risk of estrogen-related breast cancers. Testosterone can decrease "good" HDL levels and boost "bad" LDL levels of cholesterol. It can also increase the risk of heart disease or stroke. And by stimulating the growth of prostate cells, it may increase the risk of prostate cancer.

Tinkering with LHRS hormones, in other words, is a gamble. For some people, the chance of regaining the vigor and virility of the reproductive life stage and possibly extending the life span is worth the risks. On the other hand, there are some advantages to embracing the transition into the post-reproductive period, and understanding the logic of the intelligence system may at least temper some of the apparent disadvantages.

For example, psychological problems like depression—although attributed directly to diminished levels of estrogen at menopause—are more generally associated with the passage from the reproductive to the post-reproductive period of life. It is a transition that has many of the same characteristics as the often-troubled passage from pre- to post-pubescent life stages that we described in Chapter 4. At the top of the list is an identity crisis. The self you constructed to navigate through a life stage, the skills you so diligently acquired to smooth your passage, the goals you identified, suddenly no longer seem appropriate.

When you were prepubescent, what were your goals? For most of us, the agenda was simple: be the apple of your daddy's eye; be a dutiful child; get good grades. The payoff was unconditional parental investment. When you transitioned into the post-pubescent life stage, into a larger biological marketplace densely populated by your peers (who were undergoing a similar hormonal catharsis), your goals—driven by your LHRS—shifted: be popular; be perceived as attractive; be perceived as a potential mate, and so on. Being the apple of your daddy's eye—a behavior that by now you excelled in—probably cut little ice in the rough-and-tumble world of adolescence. Your existing adaptive representational networks—the database that had helped you navigate this life stage—were no longer so relevant. And so your intelligence system adapted, and you learned behaviors that made you a viable player in this new phase of your life.

In the reproductive life stage, your goals shifted once more: be a mate; be a parent; climb the corporate ladder. Again, your intelligence system adapted: You learned the ropes and developed new skills. The last major transition is from the reproductive to the post-reproductive period of your life–perhaps the most difficult passage of all. Planning– negotiating the next phase of your life–utilizes memories from the past as the building blocks to model your future. And the past is the past. What you were once capable of is no longer any reliable indication of what you might be capable of in the future. The game has changed. Your LHRS has relaxed its powerful influence on your future, leaving you–what? Adrift? Or liberated? That all depends on your unique mind.

7

# HOW THE MIND REALLY WORKS

There's another, distant garden to visit on our tour of biological marketplaces—a garden that will help put into perspective the various elements of your intelligence system that we've already described. It's far removed from the pastoral prospect we began with in Chapter 1, on our lazy summer afternoon in the company of bees and flowers and hummingbirds. In fact, from where you are sitting now, you might imagine that you have been marooned on some infernal and primordial island. Stretching in front of you—lapped by the waves of the Pacific Ocean—are vast, folded swaths of black basaltic lava, pocked and cracked and almost burnished like coals in the blistering, equatorial sun. Their surface is covered by stunted, sunburned brushwood, showing little sign of life. Listen for a moment, and you'll hear the sounds that dominate this place: the reptilian hiss of iguanas, and giant tortoises lumbering on their ponderous excursions.

As you may have guessed, you are in the Galápagos, an archipelago of volcanic islands about six hundred miles off the coast of Ecuador. (Galápago means "tortoise" in Spanish.) To an evolutionary biologist, the Galápagos are the closest thing to a scientific shrine.

Here, on the morning of September 17, 1835, ship's naturalist Charles Darwin and several members of the crew of HMS *Beagle* rowed ashore to the island of San Cristóbal. Darwin was twenty-six years old. Almost another quarter-century would pass before the publication of *The Origin of Species* on November 24, 1859—but it was in this garden of the mind that the seeds were planted that led him to one of the most influential ideas in human history: evolution by natural selection.

Central to the idea is the concept of adaptation that we mentioned in Chapter 1. Remember natural selection's greatest hits? Opposable thumbs specialized for grasping, hearts specialized for pumping, and so on? These are adaptations, shaped by natural selection into functional devices—complex pieces of machinery so well-suited to their tasks that they could have come about only by a competitive, trial-and-error process that left the unsuccessful versions lying in the dust of history like evolutionary Edsels.

The Galápagos were crucial to the development of Darwin's theories because they constituted a kind of natural laboratory—what he called "a little world in itself." It was a world of creatures that had distant ancestors on the mainland but had subsequently been so transformed by their island habitats that they constituted new species—with novel adaptations that underpinned their success.

The textbook example is the case of the birds now known as Darwin's finches. (Ironically, Darwin didn't initially see the significance of "his" finches. In fact—although he was a brilliant field naturalist—he rather botched the collection of finches, incorrectly labeling some of his specimens. It took an expert ornithologist to untangle the mess. So although popular accounts tend to portray Darwin's experience in the Galápagos as a kind of eureka! moment—something akin to [and just as misleading as] the apocryphal story of the apple falling on Newton's head, it wasn't that simple: The idea crept up on him more slowly.)

There are thirteen species of finch on the Galápagos, each with a different beak. Some are broad and strong, powerful enough to crack the toughest seeds. Others are narrower, longer, and more pointed, capable of probing into cacti. Some of the finches eat grubs; others are vegetarian. Some have beaks that can strip the bark off twigs; others have beaks that can clean ticks off iguanas' backs as the finches hitch a ride. A beak for every finch and a job for every beak. That is, unless the economic climate changes—which it can and does, sometimes with alarming rapidity. The gardens we've visited may seem, at first sight, to be quite static. In fact, they can be places of considerable flux.

Princeton University evolutionary biologists Peter and Rosemary Grant have spent more than two decades documenting the life and times of Darwin's finches (Darwin, by the way, spent only five weeks in the Galápagos) and have shown evolution in action. A long drought, for example, radically changes the dynamics of the biological marketplace where the finches ply their trade. The finches best-equipped to survive a drought are those with big bodies and big beaks, powerful enough to tear into the pitifully small remaining supply of tough seeds. This hard tack diet allows them to eke out an existence until the next rains. Other finches suddenly find they have the wrong tools for the job. There are no grubs to skewer on the end of a cactus spine held in a slender beak, so those finches perish. In nature's economy, redundancy is death. And so the population changes.

Adaptations are heritable—that's why you have some characteristics that resemble those of your parents and their kin—so the big-beak variation should be preferentially passed on to the next generation of finches. And that is exactly what the Grants found. The rains came, and the drought survivors bred, passing on their genetic recipe for success and siring a league of linebacker-sized finches with beaks to match. Eventually, over generations, adaptations become species-typical—universally found in all members of the species. The good news is that this serves to "lock in" to the genome a tried and trusted design; the bad news is that a change in the environment can transform an adaptation into a death sentence.

In the years following the drought, for example, the Grants documented evolutionary change in the finches after an extraordinarily wet El Niño episode, when seeds—particularly smaller seeds—were abundant. The linebacker finches were now at a disadvantage. Their skill was cracking big, tough seeds, which were harder to find in the sudden cornucopia of smaller types. And although they could eat the small seeds, they needed many more to sustain them. And so the story reversed itself, with linebacker finches being more vulnerable and other species of finch becoming more prolific.

So far, we've been talking about structural adaptations. But now, instead of thinking about bodies, think about brains—and minds. If bodies are dynamic systems of complex adaptations—hearts, lungs, wings, beaks, gills, opposable thumbs, and so on—it makes evolutionary sense that brains, too, have adaptations. For example, everyone has a suprachiasmatic nucleus (SCN)—the essential component of the human biological clock. Everyone has a hypothalamus, essential for maintaining homeostasis. Everyone has basal ganglia, selecting and sequencing thoughts and behavior. Everyone has a life history regulating system (LHRS), scheduling transitions through life stages. These components of your intelligence system are adaptations. In your genome, there are instructions for making an SCN, a hypothalamus, and so on.

Remember that these adaptations are the product of a long evolutionary history in which the basic *function* has been conserved. For example, you can find some kind of time-keeping mechanism in yeast, ferns, fruit flies, and crabs. All these species have genes that direct the manufacture of proteins that oscillate in time with celestial rhythms—the rising and setting of the sun, the ebb and flow of tides. From the beginning of time, it has been essential for all organisms to be in sync with the world in which they evolved. Clock genes give rise to time-keeping adaptations that coordinate all life activities, setting the beat for behavior. These adaptations are venerable instincts.

So it is with all the subcortical structures—the components of the LHRS, the basal ganglia, the monoaminergic systems. They are part of your inheritance. (This doesn't mean, incidentally, that we all have *identical* architectures at the subcortical level. Broadly speaking, we're very similar—as you might expect from the selection pressures that create species-typical adaptations. But just as no two hearts are exactly the same, neither are two basal ganglia systems.)

As you know by now, there's one more component to your intelligence system: your neocortex. And this is where things get interesting. This is the "wonder tissue" that allows you to create representations, and—in conjunction with the other elements of your intelligence system—to build adaptive representational networks.

You'll recall that your behavioral intelligence system associatively links sets of information into these adaptive units. Four components of information comprise an ARN: internal state, environmental stimulus, behavioral information, and outcome information. All of this information reflects the influence of natural selection. Its character has been largely determined by evolved sensory, motor, and internal control systems, as well as those involved in its construction, global calibration, and life history modulation.

First of all, you are in some sort of stimulus environment that is generating inputs to your sensory system. Then the activation levels of your various subcortical adaptations—like the hypothalamus—form a representation of your internal state. (Are you hot, hungry, thirsty, afraid, angry, sexually aroused?) That defines the problem that your intelligence system needs to solve. Third, you generate a behavior; and finally, there's an outcome, which is then assessed in terms of its adaptive value by your LHRS components. (Do I now feel cooler, sated, calm? Did life get better? Worse? Stay the same?) Each time, as we said, the time-tagged ARN then enters the database that constitutes the record of your trajectory through life.

We're reminding you of the architecture of these networks to underscore a critical point: They are always acting in an adaptive manner—they're adaptively constrained, to use the language of evolutionary biology. What does that mean? Just that these networks are constructed according to the rules of natural selection. The design of their construction is a consequence *of* natural selection. The system that creates these networks in you exists because your ancestors—who also had such a system—were more successful and passed on the winning recipe in the genome.

Typically, we think of a constraint as something limiting, something that sets boundaries on what is possible. And that's about right. The laws of physics and the environments in which our ancestors evolved have constrained our sensory systems, for example, so that we can see unaided only in the visible spectrum (light of 400 to 700 nanometers) and hear unaided only in an auditory spectrum that ranges from frequencies of 20 hertz to 15 kilohertz.

But these constraints had adaptive utility for our ancestors. They enhanced survival and reproduction in an energetically cost-effective way. For example, the constraints imposed by natural selection on our visual system eliminated the additional costs of perceiving light waves that did not affect the viability of our ancestors. Seeing in the ultraviolet bought us nothing in the markets of life. (The evolved intelligence system we have described, and the ARNs it creates, have been similarly constrained.)

You could think of an adaptive representational network as the fundamental unit of information processing in your intelligence system. But remember that it always reflects your individual circumstances. An ARN—in the abstract—

is a species-typical piece of functional machinery. But in physical, operational reality, each ARN in you, and in the entire population, is unique. Only you have *your* ARN database—which both connects you to and yet sets you apart from everyone else.

## HOW YOUR MIND *DOESN'T* WORK

We didn't always think this way, incidentally. For many years, we subscribed to a school of thought that is often called evolutionary psychology (or EP). It was a first-pass attempt to apply Darwinian principles to the making of the mind. Darwin himself had hinted at it in a tantalizing passage toward the end of *The Origin of Species.* "In the distant future," he wrote, "I see open fields for far more important researches ... Psychology will be based on a new foundation. . . . Light will be thrown on the origin of man and his history."

The basic EP idea was beguilingly simple: If the body is a collection of adaptations produced by natural selection, then perhaps the same proposition applies to our brains and minds. If there are specialized bodily organs—like hearts that pump blood and livers that remove toxins—perhaps there are also specialized "mental organs." And just as adaptations like hearts and livers and beaks were locked into an organism's evolutionary toolkit, written into its genome, perhaps mental organs were, too.

Of course, bodily organs have a tremendously long evolutionary history. The heart that is beating inside you as you read this has been rigorously bench-tested; so has the opposable thumb that you use to turn this page. But what about your mental organs? The EP idea is that the neural circuits that make up your mind were "designed" by natural selection to solve the problems encountered by generation after generation of your hunter-gatherer ancestors living in the Pleistocene–the Stone Age–which lasted from (roughly) 2 million years ago until about 100,000 years ago.

What kind of problems? Problems related to survival and reproduction. Problems like finding a mate, finding resources, avoiding predators, negotiating social contracts, and so on.

Each of these innately pre-specified neural circuits–or mental modules–was thought of as a kind of dedicated mini-computer, designed by natural selection to solve just one problem and operate in just one mental territory, such as the mate territory or the resource territory (the technical term is *domain-specific*). As their research program developed, evolutionary psychologists developed a model of the mind in which the modules proliferated in an effort to explain behavior: domain-specific modules for facial recognition, for detecting cheaters, for taste preferences, for assessing the desires and motivations of other people (a "theory of mind" module), for sexual jealousy, for creating a self-concept.

The EP mind, in other words, is conceived of as a federation of hundreds–maybe even thousands–of these modules, or "instincts." This was in many ways a reaction to the model of the mind to which most social scientists, anthropologists, and cognitive psychologists subscribed. They tended to view the mind as a content-free, general-purpose learning mechanism–a modern computer equivalent of Locke's blank slate.

At the root of the debate was a fundamental question: How is knowledge acquired? Is the mind like a general-purpose computer that gets its "programming" from outside (from cultural influences)? Or is it pre-programmed with a bundle of instincts? Is it–to use another EP metaphor–like a Swiss Army knife, with different blades and tools for different specialized tasks? Or is it some combination?

Evolutionary psychologists argued that a mind designed to function as a general-purpose computer was doomed to failure, because of a brain-twister known as the frame problem.

This is one of the most fundamental problems in the representation of knowledge. Here's the issue: How much pre-specified knowledge must a system have to operate adaptively? This conundrum is a nightmarish version of the chicken-and-egg problem. To solve a problem, you need to know what is relevant and what isn't. Of course, that entails understanding the problem and having a pretty good idea about how it might be solved. But if you already knew how to solve it, it wouldn't be a problem.

A lot of this intellectual hand-wringing derives from the field of artificial intelligence. Researchers quickly discovered that trying to program a robot to do what for us are trivial tasks turned out to be extremely difficult. Symbol-processing systems are so literal that every conceivable possibility has to be explicitly spelled out. For example, suppose you are trying to build a robot that can recognize a bird. If you endow your robot with a general-purpose learning mechanism (with no innate knowledge–no "clues" about what birds are), the chances of the robot figuring out the right solution by chance are about as remote as the proverbial monkeys tapping away at typewriters and writing *Hamlet* by chance.

To give your robot a sporting chance, you could supply it with a description that represented a bird. How do you represent a bird? Well, you could say that all birds have feathers and can fly. Now at least your robot has a definition to work with—something to jump-start its cognitive processes as it searches the possible universe for birds. Defining a bird as a creature with feathers that can fly has reduced the problem-solving space.

But by how much? Has it solved the frame problem? For example, suppose someone points out that emus and ostriches are flightless birds. So now you have to revise your decision-making rule to something like: Birds have feathers and can fly, unless they are emus and ostriches. And how about your Aunt Mildred wearing her feather boa and flying from New York to Los Angeles? So you revise your definition again to say that a bird has feathers and can fly—unless it's an ostrich or an emu, or one of a class of featherless bipeds called human beings, wearing a feather boa and traveling by air. So far, so good. But what if the bird is chained to the perch, or dead, or wearing very heavy boots, or stuck in treacle, or has had its wings clipped, or. . . . and on and on.

The trouble, as you can see, is that the list can be endless and unbounded. Since every problem faced by our hunter-gatherer ancestors was equally capable of generating possibilities upon possibilities—with no constraints on what information or courses of action to exclude from their decision-making processes—they would have been paralyzed.

Let's make this less abstract. Imagine that you are a hunter-gatherer searching for food in the African savannah. What knowledge do you need to forage successfully? Or—to approach the food chain from another perspective—suppose you are foraging and suddenly disturb a lion? What do you do? The EP claim is that a general-purpose computer (in this case, your mind) that had no built-in circuitry that was already familiar with—and could solve—the foraging problem or the threat-avoidance problem would take an interminably long time to run through all the possible permutations of what was food, or what to do in the presence of a lion. Each moment—without the intervention and supervising guidance of a specialized module—the alternatives would multiply exponentially in what mathematicians call a combinatorial explosion.

The result is that you would either never eat lunch—or you would *become* lunch.

On the other hand, if you had modules that prompted you to make adaptive foraging decisions or threat-avoidance decisions, you would survive, reproduce, and pass on the genes that specified those life-saving behaviors.

So in the view of evolutionary psychologists, nature's solution to the frame problem was the creation of an enormous confederation of information-processing programs, or instincts, that generated pre-specified solutions to a range of adaptive problems. They acknowledged that there must be some kind of integrative circuitry that could select the appropriate instinct when necessary, and perhaps some limited number of domain-general information-processing systems. But the emphasis was on instincts—on the innate knowledge needed to solve the frame problem—and this was the direction their research program took: a search for the modules of the mind.

The idea of a modular *brain*—with distinct anatomical and functional areas specialized for solving adaptive problems—is by now mainstream. As a result of mapping with single electrodes and imaging techniques like functional MRI, cognitive neuroscientists have known for some time that different sectors of the visual cortex, for example, process different kinds of information. It's as if the final image that we see is in some sense assembled as a result of efforts by various neuronal subcontractors that each have responsibility for their own parcel of the cognitive real estate. In area MT (short for middle temporal visual area), neurons are especially sensitive to movement—the direction of visual motion. If the MT subcontractors have done their job well, you perceive motion smoothly in real time. If MT is damaged, successive moments are disturbingly erratic, as if a perverse editor has willfully snipped out crucial frames from a movie. Damage to the area known as V8—the specialized domain of the color company—limits you to seeing only in shades of gray: The color literally goes out of your life.

Just as evolutionary psychology extended the concept of adaptation from physical organs to mental organs, it also extended the idea of a modular brain to a modular mind—to domain-specific cognitive circuits specialized for solving adaptive problems. On the face of it, that seems plausible. The mind, after all, is a product of the brain. This intellectual marriage—of adaptationism and modularity—is perhaps the defining characteristic of the EP enterprise.

But does it stand up to scrutiny? It sounds fine at a theoretical level, but mental life is ultimately conducted by neurons. The reality of how they process information (from the simplest cascade of chemical neurotransmitters to the capacity we have to conceive of relativity, fall in love, or contemplate our own deaths) constrains theoretical excesses. If an idea doesn't work in the trenches of axons and synapses—at what a neuroscientist calls the implementation level—it has no future.

With that caveat, what could it mean to have a modular mind made up of possibly hundreds of instincts—specialized circuits like a cheater-detection module, a mate-selection module, or a taste-preference module?

It seems obvious that these mental organs are not going to resemble physical organs. You can locate and transplant a heart or a liver. You can locate and transplant a subcortical adaptation like an SCN (so that the recipient animal actually takes on the rhythms of the donor). You can locate your amygdala, your basal ganglia, your hypothalamus. You can even locate, as we said, low-level processing areas of your neocortex—like MT and V8. You can also surgically move the sensory inputs from one area of mammalian cortex to another, so changing the location of the cortical representational area for that sense. Now, what about the dedicated cognitive modules of evolutionary psychology?

Do they really exist? Do you really have genetically pre-specified, reliably developing cortical circuits that predispose you to choose certain foods or be attracted by the physical characteristics of a mate or the physical characteristics of a certain landscape? (And not only you. *Everyone.*

Because if these mental organs are adaptations, they are species-typical–universally present in all our minds, whatever our culture.)

At this point, as you know from what you've read so far, we part company from the EP program. Why? Remember there were three components of their approach: a bundle of instincts, an integrative circuitry, and some limited, general-purpose computational machinery. Evolutionary psychologists bet the store on a hunt for instincts and essentially dismissed the other two elements as support acts for the star turn. In our view, EP had it exactly backward.

In a moment, we'll analyze two of their hypothesized instincts and offer a different interpretation. But first, let's revisit the concept of adaptations from the beginning of this chapter. Remember the beaks of Darwin's finches?

Let's be quite clear about this: Your brain (or, more precisely, your intelligence system) is not a beak. It's not an opposable thumb; it's not a kidney. It's a unique configuration of adaptations that form an integrated guidance system that *self-adapts*, modifying itself on-line as a function of your experience in the environment.

We agree that your brain is composed of neural adaptations that resulted from evolution (and the mind, remember, is a product of the activity in the brain). But these adaptations did *not* take the form of well-defined, inherited information-processing circuits that were designed to generate predetermined adaptive solutions to Stone Age problems. Rather, they took the form of components of a system that could construct adaptive information-processing networks–individualized circuitries that generated behavioral solutions that precisely fit the specific environmental conditions, bioenergetic needs, personal experiences, and unique life history of an individual.

Anyway, here's the first example. It's almost an EP mantra that we all have inherited taste preferences for fat, sugars, and salt. The story they tell is simple. During the Pleistocene, our hunter-gatherer ancestors needed moderate amounts of these nutrients to stay healthy. At the time, they were in relatively short supply: There were no fast-food outlets or supermarkets in the savannah. Animals were lean, salt licks were few and far between, and berries were seasonal. The EP position is that anyone with an innate taste preference for these nutrients would be more likely to seek them out, have a better chance of survival and reproduction, and pass on their genetic recipe for taste preferences to their offspring.

Now, of course, we're hunter-gatherers in Levi's and pinstriped suits shopping in a suburban savannah: We have access to excess. The genetic taste predispositions that once guided our diets have become potentially lethal (high cholesterol levels, increased risk of heart disease, and so on).

Let's assume that the environmental details are broadly correct. Does it follow that we now have evolved adaptations for seeking out fat, sugars, and salt? Are Weight Watchers and Jenny Craig and fat farms the result of some gross mismatch between our genetic legacy and a society of conspicuous consumption?

We don't think so. Let's acknowledge the obvious fact that, generally speaking, we humans do have psychological preferences for fat, sugars, and salt. Does that make them "instincts"? Did we inherit information-processing circuits that compel us to prefer these nutrients? (Remember, if it's an adaptation, that means virtually everyone in every culture.) Again, we don't think so.

What we inherited—all of us—is an intelligence system that continuously registers whether or not the adaptive problems of acquiring adequate nutrition are being solved throughout the changing passages of our life histories. We have fatty acid, amino acid, and glucose receptors that are constantly sending status reports to the hypothalamus (which is, as you recall, in close communication with other bodily systems). We have taste receptors (taste buds) and chemoreceptors that help identify nutritious foods and red-flag potentially toxic substances (This tastes bitter! Do you really want to swallow this?). Tastes are yoked to life history stages, which explains food aversions in pregnancy. Metabolisms also change throughout life. In bioenergetic terms, a glucose intake that was barely adequate for a college athlete can produce diabetes in later years.

These regulatory components impose "adaptive constraint" on the construction of the neural networks/information processing circuits that support our specific taste preferences, which are always changing in an on-line fashion as a function of experience.

Remember how adaptive representational networks are built? Here's the story for taste preferences. Suppose you are hungry (in the appropriate physiological state), and you taste something (have a specific sensory experience), and swallow it (produce a behavior), and the effect registers with your LHRS as having increased your bioenergetic viability (the behavior produces the desired outcome), then your brain will construct or strengthen a circuit that promotes a preference for that food.

The regulatory components act like guidelines for how taste-preference circuits should be constructed, to ensure that required bioenergetic resources will be consumed, and keep physiological measures in a range compatible with health. Unlike inherited information-processing circuits, the guidelines provided by the adaptive components of the system allow for enormous flexibility in expressed taste preferences. Humans will even eat dirt if that's the "foodstuff" in which required nutrients are to be found in a particular environment. It's called *geophagy,* incidentally, and the practice has been noted in many traditional societies, especially among pregnant women. It turns out, for example, that the soils sold in Ghanaian markets to pregnant African women are actually richer in iron and copper than over-the-counter dietary supplement pills. Does this mean that pregnant women have a heritable mental organ that predisposes them to seek out iron and copper? Or dirt? Of course not. It means they have heritable adaptations, components of an intelligence system that—among other things—monitors their nutritional status and guides their food preferences.

When do you begin to assemble the adaptive representational networks that store your database of food/taste preferences? From the moment you suckle your mother's milk, which is rich in fats (lipids) and sugar (lactose), although the precise composition varies as a result of her diet. (The process actually begins in the uterus, where you are sampling your mother's nutrients carried in the placental blood supply.)

As you grow older, your taste database grows, and your preferences change as a function of experience. It has always been this way. Imagine an ancestral hunter-gatherer child exposed for the first time to the landscape where his mother forages with other women and their children. How does he know which plants are nutritious? He doesn't *know,* but he observes. He learns. He sees that his mother doesn't usually harvest dry twigs, dead grass, and thorns as foodstuffs. Instead, she collects roots and tubers, fruits and berries. He sees that she favors some fruits and rejects others. Probably she explains why—look at the color, feel the ripeness, taste the sweetness. This is good fruit! And inevitably, he experiments. There are false starts as he discovers the intricacies of foraging. His lips pucker as he samples sour fruit. Red doesn't always mean ripe; green doesn't always mean unripe. The outcomes of these experiments—some provided by his taste buds and other physiological receptors, some by his mother and others in his social environment—shape his preferences. Eventually, he creates a personal taxonomy of tastes.

And it is personal. There's an enormous difference between the idea of an inherited instinct or mental organ that directs taste preferences and an evolved system that adaptively constrains the building of taste-preference circuits. It's a distinction that becomes increasingly important as we progress from trying to understand the psychological architecture underlying low-level psycho-physiological phenomena such as taste preferences—something common to all mammals—to attempting to fathom highly complex and uniquely human psychological and behavioral phenomena such as high-level cognitive abilities.

## NEURAL NETWORKS OF BEAUTY AND SYMMETRY

We promised you our thoughts on another favorite EP example—a possible instinct for detecting facial attractiveness. Usually, in popular treatments of the subject, headline writers tend to reduce this to "The Biology of Beauty." At first blush, this has a strange ring to it. Beauty, after all, is thought to be in the eye of the beholder—something immensely personal, varying from culture to culture, and certainly not in the purview of your genome. If there is any international gold standard of beauty, we tend to think that it is largely the result of the insidious influence of the global media: the Baywatching of the world. How could there possibly be some heritable yardstick for measuring attractiveness?

At the heart of the EP story—just as with the fat, sugar, salt example—is a simple proposition. If you were an ancestral hunter-gatherer looking for Mr. (or Ms.) Goodgene—someone whose exterior characteristics might somehow give you a clue to their potential as a prime partner for leveraging your assorted genes into the future—what would you be looking for? More precisely, what physical cues might be detected by your mate-selection radar, generating an arousal response, and prompting you to search your mental files for a pickup line? High on the list is symmetry.

In the lexicon of genetics, symmetry is a synonym for fitness—for health, for good genes, for a vigorous immune system. Symmetry is evidence of an organism whose developmental program was unimpaired by the predations of pathogens (also known as "parasite load"). Everything turned out just so: as on the left, so on the right. This has some sup-

port in the scientific literature. For example, if you manipu-
late the length of a bird's tail feathers, patching on a false extension so that one side is longer than the other, the lopsided
bird will lose out in the mating game. Female birds are
apparently dubious about its airworthiness and survival
prospects.

For a moment, let's leave the birds and go back to the
bees–the honeybees in our garden. You'll remember that
one of their many talents was the ability to zone in on flowers that were symmetrical. Studies have shown that symmetrical flowers are a reliable indication of higher yields of
nectar. Although plants don't have what we would consider
an immune system, they do have a developmental program
and defenses against predation. If those defenses are com-
promised by a heavy parasite load, their flowers–as any
gardener knows–can become stunted, discolored, and asym-
metrical. Nectar production drops, and the honeybee cir-
cling above quickly writes the plant off as a bad investment
of energetic resources.

So now we have an interesting situation. On the one
hand, there is the EP idea that we choose our mates according to the dictates (or at least unconscious Pleistocene-vintage
urgings) of an innate mental organ that equates beauty with
symmetry, with low parasite load and genetic vigor. On the
other hand, we have honeybees equating symmetry with
a square meal. Is there a conflict here? The bee brain, remember, is in many ways prototypical of our own. One
part of its intelligence system is instinctual, but it also has a
rudimentary swatch of plastic tissue that allows it to form
adaptive representational networks, tracking environmental
changes and navigating toward the goods. Is the bee making

an apian aesthetic judgment–employing a kind of innate "beauty detector"?

Do we have some type of lower-level instinctual mechanism that registers the presence of what might be thought of as perceptual primitives like symmetrical faces? It's well-known, for example, that very young infants stare longer at faces that adults have rated as attractive. And yet there is also evidence that newborns very rapidly form prototypes from the sets of faces they see immediately after birth. So what are we to make of this? Is it possible to tease apart the influences?

Since Plato's time, the degree to which something was considered "beautiful" was based on the degree to which it was symmetrical and classically representative of objects in its class. These characteristics were the essential qualities of Plato's "Ideal Forms," his philosophical conceptualization of the prototypical objects of knowledge, which were eternal and unchangeable–the quintessential "cat," "chair," "flower," and so on. He contrasted Ideal Forms with perceived objects– the "copies" of forms that our senses are privy to. Plato considered perceived objects to be imperfect because they are ever-changing, the furnishings of a world of Heraclitean flux.

There's a kind of parallel to Plato's perceived objects and Ideal Forms in the representational networks that create our operating virtual reality. As we explained earlier, the cortex forms representations of the individual objects that we perceive. Then–as a consequence of the hierarchical properties of representations–our intelligence systems form higher-order representations: abstractions of the essential features of all lower-order representations of objects in an associated class. So, for example, each time we see an apple, we generate an operating representation of that particular apple, but we also modify our higher-order representation of a proto- typical apple, which has been derived from abstracted es- sential features of the individual representations of apples we have perceived—the features that are common to all ap- ples. Our prototypical representations are "perfected aver- age" representations of objects. Extremes—variations from the mean—are smoothed out. As a consequence, these pro- totypes are symmetrical. In a sense, they are neural Ideal Forms.

In recent years, researchers have begun to explore the "averaging" nature of our psychological assessments of at- tractiveness. In most experiments, subjects are shown a se- ries of photographs (faces) of same-sexed individuals and asked to rate their attractiveness. Invariably, the winner of this beauty contest is someone who doesn't exist—a proto- typical face produced by computer-averaging the faces of all the other contestants. At first, this sounds counterintuitive. Why would the average be perceived as more attractive? Because the extremes have been removed, and the result reflects optimal shape and symmetry: the prototype.

Then the researchers took the investigation a step fur- ther. They conducted a similar study but created another vir- tual woman using the photographs of those women whose previous ratings had placed them in the top half of the at- tractiveness rank ordering. The photograph of this "better- than-average average" image was then included in the group of original photographs, along with the "average-average" image and a "less-than-average average" image. Not surpris- ingly, the "better-than-average average" image was assessed to be the most attractive. (They followed the same procedure with men's faces, incidentally, and got comparable results.)

Again: why? Let's revisit the functional characteristics of adaptive representational networks–particularly in the way they support our ability to generalize and discriminate. *Generalization* is a term that behavioral psychologist B.F. Skinner uses to describe the tendency of an animal that had learned to associate a specific stimulus with the opportunity to earn a reward then to respond in much the same way to other similar stimuli. With time and experience, animals– including humans–learn the precise stimulus situations that provide the opportunity to earn rewards and stop respond- ing in inappropriate stimulus situations. This ability Skinner calls *discrimination*.

Imagine that you are an infant just beginning to notice differences in people. As your cortex develops, and you gain experience in the world, you develop representations of individuals, both male and female. Then, on the basis of sub- tle and not-so-subtle differences between them, your intelli- gence system creates distinctive prototypical representations of each sex that enable you to discriminate between them. In the real world, of course, males and females fall somewhere on a continuum when you consider each of the various physical characteristics that distinguish them. Even as an adult, you might walk down the street, see a person ap- proaching you, and be uncertain about his or her sex. Your ARNs help you to determine whether you're looking at a man or a woman by analyzing the stimulus features you are perceiving and automatically activating the prototypical rep- resentation that is most similar.

For obvious reasons, the ability to discriminate between male and female members of your species is a critical one: Reproductive success depends on it. But that's just one example. Let's go back to discrimination and see how it applies to the face-averaging studies we mentioned. Discrimination is a learning process. Discriminating between two similar stimuli takes experience in differentiating outcomes. We initially sort stimuli into what you might think of as mental "bins" that are broad and overlapping. When we are differentially rewarded or punished on the basis of our ability to identify a stimulus as belonging to one bin or the other, our choices tend to be at the outer extremes. Why would we go wider of the mark?

Here's an example. Assume that you are learning to distinguish between two notes of music that are very close together in pitch. The two notes are played for you and identified as X and Y. Then they are played again, embedded in a random sequence of other notes. Your task is to respond to the note Y–but not X. If you get it right, you're rewarded; if you mistakenly choose X, you are chastised.

Which note do you *initially* respond most vigorously to? You might expect the answer to be Y. But, in fact, it's actually Z–a note that is further away from the "wrong" choice. As usual, your system is playing it "on the safe side" during these early learning trials.

Now, let's return to beautiful faces. The "better-than-average average" female face (like the Z note) is perceived as more attractive than the "average" female face (the Y note) because its representation is at the extreme end of the female "bin" in our minds. Are these choices mediated by evolved beauty-detection modules? Not at all. They are formed, on-line, by the same evolved system that enables you to distinguish between any two features of the world that are structurally similar but *functionally* distinct—in that they each yield a very different outcome.

We mentioned earlier that honeybees have something akin to a sense of aesthetics. Bees haven't read Plato and know nothing about Ideal Forms. And yet they apparently "know" about symmetry. How? Imagine that you're a bee patrolling the flower beds. On your first trip, suppose you land on, let's say, twenty flowers and collect nectar. Your rudimentary intelligence system is constantly performing a cost/benefit analysis. You know that some of the flowers reliably delivered the goods—a higher nectar concentration. And you have constructed ARNs that memorialize this correlation. It turns out there's a connection between flower shape—what we call symmetry—and nectar concentration. On your next few missions, the trend is confirmed. And so you begin to favor symmetrical flowers. You construct a prototype representation of a best-bet flower: an Ideal Form. And that becomes the standard you use to assess any new flowers you encounter. The closer they resemble the prototype, the more likely you are to land on them. (At the extreme end of the symmetry "bin" in your bee mind is a representation of an almost perfectly symmetrical flower—and that is what you aim to select: an almost perfect lunch.) With each flight you refine—and redefine—the prototype, and your sampling gets more discriminating.

Like our taste preferences, our mate preferences are created by our evolved intelligence system. And our appreciation of symmetrical faces—like our (and the bee's) appreciation of symmetrical fruits and flowers—reflects the intrinsic symmetry of prototype representations: the system's higher-order, composite average of the essential features of individual representations. Our intelligence system automatically creates these symmetrical prototypes of all the representations of the environment that we hold in our minds. Detecting symmetry is not a problem that a "mate-selection module" evolved to solve. Rather, it's a general capacity of your intelligence system.

Some final points to consider. In the beauty experiments we described, the "winner" of the contest was an abstract creation. She didn't exist outside a computer. But suppose she did? Would you have chosen her? Or would you have perhaps chosen the contestant with eyes like your first lover's, or cheekbones like your mother's? In the real world, in your real life, your unique intelligence system might even veto physical beauty as a critical factor in determining your overall attraction to someone. Perhaps you're more influenced by someone's kindness, intelligence, or talents—or their love for you. There are no evolved "mate preference" mechanisms, no evolved "taste preference" mechanisms in our minds. The preferences you have cultivated are as unique as your sense of your self. They are the creation of a unique intelligence system that belongs to only you.

# 8

## GLIMPSES OF THE SYSTEM

The Charles Darwin who came aboard HMS *Beagle* in 1831 was a novice scientist. Only months earlier, he had not a clue about his future prospects. He had decided not to follow the family tradition of becoming a physician after being sickened by the sight of an operation performed without anesthesia. For a long while, it seemed that he might be destined for the Church—probably as the vicar of a country parish where he could spend part of his time attending to God's business, and the rest indulging his interests in natural history. And then—largely as a result of chance and social connections (for he was not the first choice)—he landed the job of ship's naturalist on the *Beagle*. His qualifications were minimal—he was, after all, just a newly minted Cambridge graduate with a working knowledge of geology, a fondness for beetles, and an engaging personality. Four years later, when he went ashore in the Galápagos, he was more seasoned but still learning on the job (and still, occasionally, bungling the collection and labeling of specimens—even the finches that would eventually bear his name).

But now fast-forward to 1882, the last year of Darwin's life. In the space of half a century, an extraordinary transformation has occurred: a revolution. We're not talking about just the momentous publication of *The Origin* in 1859, which set out the notion of evolution by natural selection. In the years after he returned from the Galápagos, Darwin wrote a series of remarkable notes—insights about memory, dreams (including his own), instincts, the songs of birds, the facial expressions of monkeys, the biology of blushing, the origins of taste preferences, the emotions, music, and much more.

Liberated from the intense five years of collecting and annotating that marked the voyage of the *Beagle,* his synthetic mind could now begin to sift meaning from the accumulation of data. And it is at this point that he began to explain the implications of his discoveries for psychology: for the mind. Methodically, in a series of seminal books and papers—*The Descent of Man* (1871); *The Expression of the Emotions in Man and Animals* (1872); *A Biographical Sketch of an Infant* (1877)—he laid the foundations for an evolutionary approach to comparative and developmental psychology.

Virtually every day since moving there in 1842, he took a turn or two around the Sandwalk—he called it his "thinking path"—that skirted the edge of his home, Down House, in the countryside south of London. There, surrounded by nature, he reflected on human nature. (You can almost hear him now—the rhythmic click, click of his iron-tipped cane biting into the gravel as he paced through the copse of trees, past the sloping lawns and brilliant colors of the flower beds.)

Forty years and more of pondering where we came from, why we behave as we do. Forty years that forever changed the way we think about ourselves.

Darwin's influence on psychology has been pervasive. You can see it in the work of William James (his *Principles of Psychology* was published in 1890), of Sigmund Freud, of B.F. Skinner, of Abraham Maslow, and more recently, the current group of evolutionary psychologists. The theory of natural selection had a major impact on these thinkers and their models of the mind. But none of them had understood the way in which fundamental energetic principles drove the evolution of life intelligence systems. And so their theories reflected only a part of the picture.

There is a parable of the blind Indian sages who each grasped a piece of the elephant but missed the whole. So it has been for the principal system builders of twentieth century psychology: psychoanalysis, behaviorism, the "third force" of transpersonal psychology, even EP—each of them has a piece of the elephant of human nature. How do these other theories map onto the model we've described? Does our version encompass the still-salient pieces of schemes that have in many ways been discredited? We dealt with EP in the last chapter, so let's take a look at Freud's legacy.

Freud was a young man of almost twenty-six when Darwin died in April 1882, and died himself shortly after the outbreak of World War II in September 1939. Again, in the space of another half-century, there was a revolution in how we perceived ourselves. It was evident in everyday language. Perhaps someone was anally retentive, or repressed. Perhaps they were being defensive, or off on an ego trip. Maybe they made a Freudian slip. Did a boy suffer from an Oedipal complex? Did a girl have penis envy? "If often he was wrong and at times absurd," W.H. Auden wrote about Freud, "To us he is no more a person/now but a whole climate of opinion/under whom we conduct our different lives."

Of course, to have created a climate of opinion doesn't mean that the legacy was scientific. And that's always the primary criticism leveled at Freud and his brainchild, psychoanalysis. Ironically, Freud began his career as a scientist. He did creditable research in neuroanatomy and later in clinical neurology. It is apparent from his *Project for a Scientific Psychology* (1895) that that is where his ambitions initially lay. But exploring the mind scientifically at a time when there were virtually no methods to conduct research proved fruitless, and the *Project* was eventually abandoned stillborn. Freud instead tried to understand the mind by exploring the murky world of the unconscious, panning for psychic gold in the free associations that he prompted in his patients.

This foundational technique of psychoanalysis—following a kind of paper chase of buried ideas—originated with Freud's colleague Joseph Breuer. In 1880, Breuer began to treat a patient he called Anna O. (To disguise identities in his case notes, Breuer devised a simple code, based on a one-letter alphabetical shift. Thus the real Bertha Pappenheim—Bertha P.–became Anna O.) Breuer discovered that, during hypnosis, he was able to provoke a trail of mental (and therefore physiological, emotional) associations in Anna O.

What these pioneering psychoanalysts had stumbled upon was the associative nature of adaptive representational networks—the functional fabric of the mind. They used it to develop the technique of free association—Anna O. called it the "talking cure."

There is a huge literature on these early days of psychoanalysis that we don't need to explore here—including criticism of Freud for downplaying the fact that Anna was not "cured." Even so: The key point is this. Freud saw utility in the idea. He fine-tuned it, abandoning the need for hypnosis, and simplified the technique to the scene we all recognize: "requiring the patient to lie upon the sofa while I sat behind him, seeing him, but not seen myself."

How could this simple probing possibly work? How, as Freud asked in his autobiographical study, could it possibly be that his patients "had forgotten so many of the facts of their external and internal lives but could nevertheless recollect them if a particular technique was applied"? His answer was that "Everything that had been forgotten had in some way or another been distressing; it had been either alarming or painful or shameful by the standards of the subject's personality. It was impossible not to conclude that that was precisely why it had been forgotten—that is, why it had not remained conscious. In order to make it conscious again in spite of this, it was necessary to overcome something that fought against one in the patient; it was necessary to make efforts on one's own part so as to urge and compel him to remember. The amount of effort required of the physician varied in different cases; it increased in direct proportion to the difficulty of what had to be remembered. The expenditure of force on the part of the physician was evidently the measure of a *resistance* on the part of the patient. . . . I was in possession of the theory of *repression*."

These are some of the key components of Freud's system. At the core is the idea of the unconscious. This was not a discovery of Freud's: Others (including William James) had talked about the unconscious as a kind of vast, subliminal warehouse of information that could be supplied, on demand, to the conscious mind. Freud's conception was more dramatic. He saw it as a kind of psychic cauldron, seething beneath the surface. In this volcanic metaphor of the mind, unacceptable, anxiety-provoking wishes and drives were repressed until the pressure became intolerable and they erupted into consciousness.

This Freudian idea—that we are unaware of much of what motivates us and drives our behavior—is now part of our received wisdom: We accept that we are not privy to the gossip of neurons underlying our decisions. But in Freud's day, it was a revolutionary proposal.

How did Freud imagine solving the problem of repression—tapping into and releasing the psychic "pressure"? Through free association—guiding a patient along a series of mental stepping-stones, down the pathway of her past to . . . what? To some moment, some event (almost certainly from childhood), buried deep in the recesses of her mind that—once recognized and understood—would discharge the pressure, release the pent-up energy. (If this sounds rather quaint, remember that these hydraulic images were as much a part of the physics of Freud's day as the computer metaphor of the mind is of ours.)

In a moment, we'll show you how all this makes sense in the language of adaptive representational networks. But first, let's go back to Darwin.

We mentioned that Darwin had a tremendous influence on Freud's intellectual development. In the early 1870s, while still at Gymnasium (roughly equivalent to high school), Freud recalled later, "the theories of Darwin . . . strongly attracted me, for they held out hopes of an extraordinary advance in our understanding of the world."

We know, too, that Freud owned copies of many of Darwin's books. Did he perhaps read Darwin's autobiography, published in 1887? Because there he would have found a tantalizing, almost throwaway passage that might have made him wish that Darwin were still alive and could be tempted to spend time on the couch.

"My mother died in July 1817, when I was a little over eight years old," Darwin wrote, "and it is odd that I can remember hardly anything about her except her death-bed, her black velvet gown, and her curiously constructed worktable. I believe that my forgetfulness is partly due to my sisters, owing to their great grief, never being able to speak about her or mention her name; and partly to her previous invalid state."

What would Freud have made of an eight-year-old with such sketchy memories of his mother? What would he have made of an astonishing letter that Darwin wrote to an old university friend whose wife had just died after a long illness? "I truly sympathise with you," Darwin wrote, "*though never in my life having lost one near relative* [italics added], I daresay I cannot imagine how severe grief as yours must be." And what might Freud have thought of a man who suffered through much of his adult life with a menu of maladies ranging from hypochondria and chronic digestive problems to anxiety attacks and depression?

We don't have Freud to ask, but we do have some clues in the British psychiatrist John Bowlby's biography of Darwin. Bowlby, who died in 1990, was celebrated for his "attachment theory," in which he explained the crucial importance

of the early attachment of a newborn to its primary caregiver. Attachment theory drew upon two fields—evolutionary
biology and psychoanalysis—essentially weaving together
Darwinian adaptationism and Freudian ideas of personality
development. So Bowlby was unusually well placed to perform a kind of armchair analysis of Darwin's character. His
verdict was that Darwin's loss of his mother, coupled with a
lack of opportunity for mourning, were in large part responsible for his fragile constitution. In Freudian terms, he had
repressed the memory of his mother, producing the psychic
"pressure" we mentioned earlier, which led inexorably to
the eruption of Darwin's various illnesses.

In Freudian terms, repression is one of a number of
defense mechanisms we use to bury unacceptable material
beneath the surface of consciousness. Freud viewed psychoanalysis as a kind of mental archaeology, in which these
repressed memories are exhumed from the deepest strata
of the psyche. The more deeply they are buried, the more
severe the anxieties and neuroses, and the more protracted
the treatment. Our fictitious encounter between Freud and
Darwin, for example, could have been a long, drawn-out
business. But more than likely, it would have begun with the
kind of instructions from Freud that he gave to patients as he
began the technique of free association: to say spontaneously,
truthfully and without censoring themselves, the first thing
that came into their mind.

Suppose, for example, that—magically—Charles Darwin
were persuaded to enter analysis. Suppose friends and
family convinced him that all the nostrums, the dietary fads,
the ice-cure, the water-cures were doing him no good and
that underlying his various illnesses was some psychological

problem. And finally, let us suppose that the root cause was Darwin's loss of his mother (Bowlby's thesis). How would Freud's "talking cure," using free associations, play out?

First think about the problem in terms of his memories—his representational networks. At the age of eight—despite his later denials—Darwin must have formed a substantial number of networks "featuring" his mother, Susannah. Think about your own experience of early childhood: vacations, birthday parties, school events, unforgettable domestic moments. Darwin must have had similar memories (and the representations that supported them). So what happened to Darwin's "Susannah" representations?

The death of his mother must have been a severe and painful jolt to his system. But what does that mean operationally? For one thing, ARNs associated with the event would have a "painful" internal-state component. Remember that in the process of mentally modeling, an alternative trial reality is being created. And in this interior reality, as in the external world, painful stimuli are avoided. Every remembrance elicits more pain and produces negative physiological effects. You feel bad. And so networks associated with pain are rejected as building blocks for models of future behavior. Your intelligence system is saying, in effect: Let's not go there. So how can a more tolerable reality be constructed?

Mourning is a process that is universally used to attenuate the pain of the loss of a loved one. Across cultures, there are rituals associated with death. Why? The rituals activate and modify memories of the deceased in a supportive social setting, where a community of mourners participate in "adding value" to those most affected by the loss. They eulogize the deceased—affirming, by association, the value of the survivors. They bring gifts: flowers, food, tokens of esteem. We often say that the funeral or the wake is as much for the living as the dead—and if you think about it in terms of the nature of your mind and the biological marketplace, that makes perfect sense. All of this serves to affirm the bereaved's future viability in the biological marketplace.

With these affirmations, the networks that represent the initial experience of loss are reactivated and updated; the network values are changed. Through the process of grieving, the memories of the loss become less painful.

It is inconceivable that Darwin did not think about the loss of his mother after her death. Some stimulus or another would have evoked memories of her, and the pain of her loss. The trouble in Darwin's case seems to be that the healing process of mourning was cut short. Almost as soon as the funeral was over, his mother's name was not mentioned again. His father became depressed and adopted a sarcastic, carping manner toward him. And his two older sisters appear to have taken their substitute-mother roles to extremes, and rather scolded young Charles.

Susannah Darwin's death was territory to be avoided. And so Charles's memories were never gracefully modified. They were sequestered in a kind of psychic lockbox in childhood—repressed, in Freudian terms. The task of analysis would be to use free association to help the adult supply the key to the lockbox—in our terms, activate and modify those ARNs.

In this imaginary analysis, it is hardly surprising that Freud might have zeroed in on childhood. Like us, he was only too aware that early experience is crucially important in determining the trajectory of a life. We view this maturation of a system in terms of energy allocation—always meeting the bottom line.

For Freud, though, the driving energy was sexual. He saw development as a psychosexual process. He believed that we inherit drives, instincts that propel us to be who we are, yet he recognized that early experiences engender adult psychology. He placed enormous emphasis on the formative years of childhood, arguing that the events of the first five or six years of life determine psychological development. As a psychiatrist, of course, Freud developed his views on the basis of his patient population's problems, rather than on an understanding of the functional design features of the mind. So it's perhaps not surprising that he would assess personality to be a static feature, for all the reasons we have described as obstacles to change (see Chapter 4).

Freud believed that all pleasure is ultimately sexual pleasure, that libido—our drive toward sexual pleasure—is at play in all the major activities of our daily lives. According to his theory of psychosexual development, we spend the first year of life in the "oral stage," a period during which the mouth is the erotic focus, and the key task to be accomplished is weaning. From the age of one to three, we enter the "anal stage," in which the erotic focus is the anus and the primary task is toilet training. Then, from three to six, we are in the "phallic stage," with the erotic focus being the genitalia and the task identifying with adult role models. Between the ages of six and twelve, we lack an erotic focus, according to Freud. He believed that we are sexually repressed during this period, which he called the "latency stage," and our primary task is to expand social contacts. Then, from the age of twelve onward, he believed, we are in the "genital stage," in which our erotic focus is the genitals, and our task is to establish a family and generate new life.

How does this square with the model we've been describing? In some ways, Freud's oral stage resonates—acquisition of bioenergetic resources is clearly the predominant force driving the earliest stages of development. But his notion that the libido dominates human motivation throughout the life span contrasts sharply with our description of the LHRS modifying the motivational state of the individual in the sequential stages of life.

Freud's model of the human mind, on the other hand, fares better. He actually developed two complementary models: a "structural" and a "topographical" model, and it is some measure of Freud's enormous influence that you'll almost certainly recognize the terminology.

His *structural* model divides the mind into three units with distinguishable functions: the *id, ego,* and *superego.* According to Freud, these subsystems of human personality come on-line following a developmental course.

*Id* is Latin for "it." The id is the bratty, self-centered part of the personality, containing all the basic biological urges: to eat, drink, eliminate, stay in a comfortable homeostatic range (of temperature, for example), and gain sexual pleasure. The id operates under the pleasure principle—demanding immediate gratification. Its motto is "I want." But because the world doesn't always care to cater to an infant's whims, the id comes prepackaged with an operative process, the *primary process.* According to Freud, if the thirsty infant doesn't get mother's milk, he creates an "identity of perception"—in this case, a mental image of milk—to reduce the tension created by the urge. Because the id makes no distinction between fantasy and reality, this *wish fulfillment* serves to satisfy the urge.

The second personality unit is the **Ego,** Latin for "I." It develops as the buffer between the id and reality, attempting to suppress the id's urges until an appropriate situation arises, or by implementing rules of conduct ("Say 'May I'!"). This repression of socially inappropriate urges represents the greatest strain on—and the most important function of—the mind. So, according to Freud, the ego utilizes defense mechanisms to facilitate it. Whereas the id meets an unfilled urge with fantasy, the ego devises a strategy to fulfill the urge in reality. Freud believed that the ego acts in service of the id, but utilizes psychic energy to control its urges until they can be satisfied in a socially acceptable manner. The ego's efforts to resolve conflict between the needs of the id and the constraints of social reality lead to the development of skills, memories, and, ultimately, self-awareness. With the development of the ego, the individual becomes a "self," instead of an undisciplined collection of needs and demands. This personality unit operates under the *reality principle,* recognizing the external world as it actually is. It employs the *secondary process,* reason, in its efforts to obtain pleasure.

The final structural model unit is the **Superego,** which is Latin for "over I." The superego is Freud's notion of a conscience, a sense of right and wrong that develops with the internalization of parental representations. While the ego merely temporarily represses the urges of the id that are likely to result in punishment, the superego internalizes the relationship between acting on the urge and punishment. The child, for example, will not steal the cookies even when unwatched because he has internalized the likely admonition. The superego uses self-reproach and guilt as its primary means of ensuring that these internalized "don't rules" are enforced. But when a person does something that the superego recognizes as socially laudable, it reinforces the action with self-satisfaction and pride.

Freud's structural model taps into a number of important features of the integrated intelligence system. The characteristics of the id, for example, reflect the core intelligence system's functional goal of ensuring that an individual's basic needs are met. Freud's notion of the pleasure principle equates perfectly with the core system's design to register acquisition of "the goods" with a positive affective sensation. Of course, it is virtually impossible to reconcile the model with Freud's contention that a fantasy good registers as the same thing as an actual good when it comes to basic bioenergetic needs: You can't quench your thirst with a drink of imaginary water. But with the construction of ARN networks in the developing cortex, sensory representations of the goods—such as a mental image of a cup of water—are being functionally associated with internal drive states, like thirst. As a child begins to navigate her environment, sensory representations have real adaptive utility: They are nature's way of actually fufilling our vital needs—and wishes.

According to Freud, the ego enables a child (and ultimately, an adult) to plan and delay gratification. It comes online when the individual has developed enough cortex to have a functional behavioral intelligence system. Something to plan *with*. Now, the child also has a representation of self. Now she *can* be self-aware.

Finally, the superego reflects the hallmark function of ARN networks—the guidance of behavior in a manner that is most likely to get the vital goods and avoid harm in the environment.

Freud's *topographical model* carves the mind into three levels of self-awareness: the conscious, preconscious, and unconscious minds. The conscious mind contains information that we are aware of at any given moment: Anything that is thought, perceived, or understood resides at this level. One level down is the preconscious, which contains the memories and thoughts that are easily recalled, ready to break into consciousness at any moment. At the bottom level is the unconscious, which contains the personal information that we are not aware of: the drives, urges, wishes, and thoughts of all of our past experience—by far the largest piece of real estate in the psyche. According to Freud, the contents of the unconscious mind threaten to destabilize the conscious mind if they surface.

The parallels between Freud's topographical model and the model of the integrated intelligence system we've described are pretty obvious. The unconscious mind corresponds to the underlying, highly self-interested motivations that characterize the most basic functional aspects of the intelligence system, as well as those memories that are associated with debilitating pain. While some of these motivations enter into consciousness (those that Freud would attribute to the "preconscious" level), others are encoded in networks with sufficient negative outcome that they do not. Remember that ARN networks guide much of our behavior in the absence of conscious awareness. As Freud intuited as he identified and categorized ego defense mechanisms such as denial, our minds protect us from being consciously aware of all our motivations. Conscious experience arises from our system's modeling of a potential reality that is likely to work for us in the world; social expressions of our self-centered or otherwise socially repugnant motivations usually don't. For good reason, our thoughts usually come to the fore dressed in their Sunday best—appropriate for our sense of our social self, as well as for the social world.

Let's sum this up before we look at another piece of the elephant. Freud's mapping of the mind and personality into these two tripartite schemes has at its core the notion of resolving psychological conflicts in the individual—conflicts that have their genesis in childhood and take place beneath the surface of consciousness. Our thoughts, our motivations, our behavior are a consequence of an ongoing interaction between our biological drives, our egoistic efforts to satisfy them (and navigate through the biological marketplace), and the moral imperatives that are brokered by the superego.

Our version goes like this. As a child, you are selfish—a stripped-down Freudian id. That may work fine when the marketplace is limited to your primary caregiver (most likely your mother). But take your bioenergetic agenda into the greater marketplace, and you may be in for a rude shock (from all the other ids). This abrupt reality check—suddenly being compelled to negotiate with other players in the market for your survival—shapes your mind and your behavioral repertoire. You have developed an ego. As you mature, you begin to form alliances with valued cooperators. Their goals become your goals. You have a superego.

Freud's models of the mind were anathema to B.F. Skinner (who thought *any* model of the mind was unnecessary). A radical behaviorist, he believed that "We do not need to try to discover what personalities, states of mind, feelings, traits of character, plans, purposes, intentions, or other perquisites of autonomous man really are in order to get on with a scientific analysis of behavior." In Skinner's view, the goal was not to understand the human psyche (whatever that was) but to show that human behavior—just like that of rats and pigeons—is entirely a response to the impact of the external environment.

Skinner's technique is called operant conditioning. Suppose, for example, you want to train a pigeon to peck at a light. How do you do it? How do you create a behavior that isn't in the pigeon's repertoire? Bit by bit, piece by piece. You begin by "catching" the pigeon as it makes a random movement that happens to be in the direction you want, and rewarding it (reinforcing the behavior). This increases the frequency of the behavior. Then you withhold the reinforcement until the pigeon, by chance, makes a new move, this time even further in the right direction. And then you reward it again. And so on. Each time, the pigeon "operates" on the environment in a way that produces some change that leads to a reward (and achieves your goal). This is operant conditioning, and animal trainers (and smart parents) use it all the time.

There are other elements to the technique, like negative reinforcement (where the learned behavior is avoidance of an unpleasant stimulus), extinction (where behaviors that are no longer reinforced gradually wane), and punishment (which needs no explanation). Skinner also worked out various reinforcement schedules (in which, for example, the time interval between reinforcements was varied). The details don't concern us here. The important point is that Skinner devised a system that consisted of building a relationship between a stimulus environment, a behavior, and the consequences of the behavior. Although he didn't think of it this way, he was building adaptive representational networks.

If you look back at our description of an ARN, you'll see that Skinner seems to have missed one element—an internal state change. In fact, it's implicit in the conditioning procedures—animals are routinely food-deprived before training. The LHRS is always monitoring and managing bioenergetic resources. If the energy intake is reduced, the system goes on alert, and there is a state change that prompts the intelligence system to solve the problem. But Skinner's focus was on the outside, not on the inside. Physiological needs weren't the primary thrust of his approach.

But they were the *essence* of the framework developed by Abraham Maslow, who imagined a hierarchy of needs that guide human thought and behavior. At the base of the hierarchy (usually depicted as a pyramid), Maslow placed *physiological needs*—securing the staples of life: oxygen, water, proteins, carbohydrates, fats, salts, vitamins, minerals, and so on—together with maintaining your pH balance and body temperature within limits; sex; and shelter (the primary goals established by the LHRS). On the next level are *safety needs*—comfort, security, freedom from fear. Then come the *social needs*—for affiliation, acceptance, friendship, a sense of belonging. The next level deals with *esteem needs*—approval and recognition (this is the domain of the ego—the self). Then, finally, there are the *self-actualization needs*, which roughly translate as trying to realize your highest potential. (As the advertising slogan for the U.S. Army put it: "Be all that you can be.")

One of the leaders of the so-called "third force" in psychology, Maslow has had an extraordinary impact on our understanding of motivation and personality development. The humanistic approach that he helped pioneer celebrates human potential rather than viewing people as conflicted and emotionally damaged, or as mindless robots.

Maslow argued that actualization is the driving force of human personality. As he put it: "A musician must make music, an artist must paint, a poet must write, if he is to be ultimately at peace with himself. What a man can be, he must be." The path to this kind of self-actualization involves climbing up the different motivation levels of the pyramid. Maslow's basic idea was that people will strive for higher-order needs—like self-esteem—only when the lower-order needs have been satisfied. A starving man wants bread, not approval or recognition. Maslow acknowledged that the hierarchy is dynamic. Needs can shift—so that, for example, a poet engrossed in self-actualizing will take a lunch break if his glucose receptors inform the other components of his integrated intelligence system that he's plummeting to the physiological-needs level.

Although the drive toward self-actualization remained at the top of Maslow's hierarchy, he later refined and differentiated the category. He added two growth needs prior to self-actualization: a *cognitive* level—the need to know, to understand, and explore; and an *aesthetic* level—the need for symmetry, order, and beauty. He also added one level beyond self-actualization: *transcendence*—the need to help others find self-fulfillment and realize their potential.

You've probably already noticed the areas where Maslow's hierarchy aligns with the intelligence system model.

Core hypothalamic needs must be met first (physiological), then limbic system/emotional needs (safety and social), then needs that rely on cortical function (esteem and self-actualization). If you look back over the earlier chapters of this book, you can trace the progression—the bacteria, the bee, and the bear satisfying their bioenergetic needs; the infant securely basking in its mother's oxytocin bliss; the developing child seeking approval, conjuring selves to navigate effectively through its social world; the adult occupying a niche in the biological marketplace.

Even so, Maslow's scheme—like much of humanistic psychology—is in many ways intuitive, anecdotal, and impressionistic. It failed to explain how the human mind moves from one level in the hierarchy to another. And like the other psychological frameworks we've discussed (and many important ones we haven't), it did not attempt to explain the greater design of the mind that gives rise to human nature.

Understanding the integrated human intelligence system, and appreciating its dynamic nature, offers renewed hope that the psychological and behavioral sciences can positively shape the future of human social interaction at every level—that of the individual, couple, family, group, nation, and international community. In that sense, Maslow's vision of self-actualizing individuals capable of transcendence takes on a new significance.

# 9

## AT ONE WITH THE UNIVERSE

"There is grandeur in this view of life," Charles Darwin wrote in the famous last sentence of *The Origin of Species,* "with its several powers, having been originally breathed into a few forms or into one; and that, whilst this planet has gone cycling on according to the fixed law of gravity, from so simple a beginning endless forms most beautiful and most wonderful have been, and are being, evolved."

*The Origin* was an instant best-seller. All the copies published on November 24, 1859, sold out that day. A second edition was rushed out and appeared in January 1860. In those few weeks, Darwin eliminated 9 sentences, added 30, and rewrote 483. One of the most significant changes he made was in that last sentence.

He added three words: "There is grandeur in this view of life, with its several powers, having been originally breathed *by the Creator* [italics added] into a few forms or into one . . ." Why the change?

This is a man, remember, who wrote in his autobiography that, during his days at university, he "... did not in the least doubt the strict and literal truth of every word in the Bible." A man married to a strict religious believer. A man who was warned by his formidable public champion, T.H. Huxley, that "he was in store for a great deal of abuse and misrepresentation ..." for transferring God's daily workload to a process called natural selection. But this is also a man who was so devastated by the death of his favorite child, Annie, at the age of ten, that his early faith in a personal God was essentially extinguished. Darwin, in other words, was deeply conflicted about the notion of a Creator–of what theologians and philosophers call Agency, a Prime Mover.

Was his ambivalence in *The Origin*–his appeasing addition of references to a Creator in the second and subsequent editions–an effort to deflect theological criticism and avoid offending his wife, Emma? Perhaps. After all, his primary objective was to win acceptance for his life's work: the theory of natural selection. And that theory stood or fell on the basis of the exhaustive evidence that he presented for the evolution of life on earth. How those powers had originally been breathed into a few forms or into one was something about which he was necessarily agnostic. He didn't know how matter came into existence or how life began. He knew nothing of the Big Bang, or black holes and expanding universes–nothing of genes or DNA, much less of God. But he *did* know how the exuberant variation of life forms could emerge from so simple a beginning. That was his battleground–the rigor of his science, not the quicksilver speculations of metaphysics.

In the intervening 150 years, scientists have unraveled many of the mysteries that, in his day, seemed insoluble. We know about genes and DNA, about the strong and weak forces of nuclear interactions. Although we are still largely baffled by the initial state of the universe (or, indeed, if there are other universes), although we still have nothing definitive to say about the existence of a Creator, we have become increasingly convinced that all natural phenomena are of a piece. Divine or not, there is what might be called a plan–a body of interrelated laws, discovered through the great communal detective story of science, that govern all forms of matter and energy in our universe and lead inexorably to the design of your mind.

In our first visit to the garden, we learned that it operated according to these laws–laws of physics, of chemistry, and of biology. Energy is the tie that binds these disciplines. The laws of thermodynamics set the limiting conditions for survival. We said that, ultimately, we all dine off sunbeams– the energetic flux that sustains, and transforms, life. The laws of chemical bonds–how they are made and broken, how the energy they contain is distributed–set the limiting conditions for the interaction between atoms, which produce molecules, which have produced life. These same energetic laws find expression in the behavior of bacteria, of flowers, of bees– and of human beings.

There is a satisfying strictness to these laws. We know, with some precision, the story of proton-proton reactions, of carbon cycles, of nucleosynthesis in stars. We know, with some precision, the energetics of the formation of amino acids, sugars, and bases–the building blocks of life. We understand the trafficking of energy by chlorophyll in plants. We know, with some precision, the terms of engagement under which two neurons can have a conversation and transmit information—terms that, in the wonderfully apt language of neuroscience, require them to be excited and produce action potentials, accurately measurable in millivolts. And we have argued in this book that your intelligence system must be governed by these same laws—must play by the rules of a bioenergetic economy.

In other words—as you know by now—any intelligence system must have been fashioned by natural selection to acquire, manage, and direct energetic resources to optimize the ability of the organism to survive and acquire the goods it needs in the moment and throughout its life span. Everything we have said in this book is predicated on this fact of life. And as we have shown you, that entails a series of conclusions about how your mind really works, and how unique individuals interact with each other at all levels, from the limited marketplace of the home to the global marketplace.

Your intelligence system wages a lifelong battle with entropy: You need to acquire energy to stay reliably in the black. (When we say "you," remember this is shorthand for the wonderfully complex interplay of hormonal, neuronal, and biochemical systems operating largely beneath the surface of your consciousness and emerging into social space.) You have to factor in changes during your life span. How do you efficiently allocate energy toward the attainment of life's successive goals—physically transforming yourself from infant to adolescent, shifting from reproductive to postreproductive years—all of this against a background of changing environmental conditions? As we've said, these tasks are under the management of your life history regulatory sys-

tem, which is responsible for scheduling and allocating resources over your life span.

You recall that there's another problem to be solved—
one that ultimately shapes and reshapes the contours of your
mind. Remember the catch-22 of energy? Behavior is expensive. You have to invest energy to get energy. How do
you know if your investment is paying off? You need a kind
of internal accountant, constantly performing a cost/benefit
analysis and letting you know if life is getting better or worse.
And all of this has to happen on-line: You can't afford to wait
for a monthly statement.

Like any responsible person keeping tabs on the family
budget, your intelligence system is keeping a record of
the energetic income and expenditure produced by your actions, making on-line adjustments, and devising new behavioral paths into the continuously uncertain future of
the biological marketplace. All your planning and decision-
making processes ultimately rely on predictions based on
this kind of cost/benefit analysis—which is, in turn, a consequence of energetic laws. That's the essence of the intelligence system we've described in this book.

The rest follows. You have a unique "database" of
adaptive representational networks (the fundamental units of
intelligence) that memorialize these changing records and
enable you to make energetically sound behavioral decisions. The ARNs give rise to self-representations, to your
various personalities, to prototypes. Because you are a quin-
tessentially social animal, dependent on your negotiative
skills to navigate through the biological marketplace, these
ARNs generate inferential circuitries that allow you to "read

minds." Your system draws on the database to model alternative realities, to generate and interpret language. And all these processes are modulated by monoamines (remember Alice?).

The patterns of your mind—of the human neurocognitive architecture—have been generated by, and faithfully reflect, the basic energetic principles of the laws of thermodynamics. And the underlying logic of your intelligence system is no different from that of *E. coli* or *Apis mellifera*. ". . . from so simple a beginning, endless forms most beautiful and most wonderful have been and are being evolved," Darwin wrote.

But consider this. If the contours of your mind are in alignment with these energetic principles, wouldn't you expect human social behavior—and the cultural endeavors of the human mind—also to reflect these patterns? Are there, in a sense, laws of the human mind that have the same authority as the laws of physics, chemistry, and biology? Laws that make us at one with the universe?

Let's visit the final garden in this book. We're at an estate called Hammarby, about ten miles from the Swedish university town of Uppsala. This was the home of Carl Linnaeus, the eighteenth-century naturalist who, in his day, catalogued what was then thought to be "Creation." In his publications—notably the *Systema Naturae*—Linnaeus classified all life as it was then known. He gave concise descriptions of each species, standardized the way they were named, and arranged them into a hierarchy of ever more inclusive groups. This was the kind of ambitious attempt at systematizing that characterized the European enlightenment—an attempt to unify all knowledge, driven by the engine of science. (Think also of Diderot and the French *philosophes* assembling their *Encyclopédie*.)

This vision of unity in diversity has its roots in what the historian of science Gerald Holton has called the Ionian Enchantment–the intellectual climate of ancient Greece in the sixth-century B.C. that continues to seduce the Western mind. The Enlightenment was a child of the Enchantment–and entertained the same belief in a world that was orderly and could be explained by a small number of laws of nature. It surfaced again in the works of the nineteenth-century Cambridge philosopher of science William Whewell, who gave it the name "consilience," and has been resurrected again at the end of the twentieth century by the Harvard entomologist Edward O. Wilson in a best-selling book that borrowed Whewell's word for its title, *Consilience*. "When we have unified enough certain knowledge," Wilson writes, "we will understand who we are and why we are here."

Let's understand the project clearly. He's not merely talking about the unification of the four fundamental forces of nature into what physicists call a Grand Unified Theory or Theory of Everything–which, heaven knows, is an undertaking of epic proportions. He's not simply talking about aligning, for example, physics and chemistry and biology and geology and atmospheric sciences–which is relatively simple. He's talking about the whole intellectual enchilada. Unifying the sciences, the arts, the humanities, the social sciences. The Big Consilience. What's needed to achieve Wilson's dream is a kind of Grand Unified Theory of the mind that flows seamlessly from the laws of the physical universe to the lives of human beings.

Linnaeus anticipated the problem. In his *Systema Na-*

*turae,* for humans (for the species *Homo*) he wrote simply: *Nosce te ipsum*–Latin for "Know thyself." Above the temple of Apollo, the ancient Greeks inscribed a similar message from the Delphic Oracle: *Gnothi seauton.* Know thyself.

It seems the simplest of tasks. Every morning we awaken, and the shards of ourselves are somehow reconstituted. A coherent being emerges to greet the new day–to make breakfast for the children, to jog a couple of miles, to smoke a pipe and write the first lines of a paper about relativity, to board a bus in Montgomery, Alabama, or a jet aircraft bound for the World Trade Center.

Our sense of self is so second-nature that we ply our trades in the marketplace with hardly a thought about the forces that animate us. We strive to meet the bottom line, not understanding that the ultimate bottom line–energy–motivates our actions and shapes our minds. We behave as if we are autonomous individuals, not always aware that our self is a consequence of our ever-changing interactions–an emergent property of our daily dealings with other players, past, present, and in our imagined futures.

Our philosophies, our religions, our laws, our economic systems, our literature, and our artistic endeavors–the entire cultural output of our species, wherever we live, whatever our race, color, or creed–are direct products of a universal design for creating unique minds. And that design is an unavoidable consequence of the energetic principles that we set out in the garden where we began this story–a story that flows seamlessly from the laws of the physical universe to the social and cultural lives of human beings.

Darwin's publication of *The Origin* in 1859 heralded a change in our worldview. A walk in the garden was no longer the same for those who embraced the principles of natural selection. They had a new lens through which to view the prodigious business of life. We believe this book offers a correspondingly novel and powerful clarifying lens—an energetic model of the mind—that will allow you to see yourself as you really are: at one with the universe.

# FURTHER READING

Here is a selected list of references that will serve as entry points to the literature we have briefly touched on in *The Origin of Minds*. For the evolution of the neocortex and a description of the gymnastics of *E. coli,* see Allman. For an alternative model of the human intelligence system, see the foundational essays in Barkow, Cosmides, and Tooby, and the pop version in Pinker. (As Chapter 7 makes clear, we fundamentally disagree with–but acknowledge the formative influence of–the EP position.) Darwin's works are still immensely readable and are enhanced by the introductory essays of Beer and of Brown and Neve; Weiner tells the modern story of Darwin's finches. Bowlby's insights about Darwin's psychology are compelling. Ramachandran's medical case studies illuminate the issue of neural plasticity. Etcoff deals with symmetry and the biology of beauty. Damasio writes about feelings; James discusses the social self. We are sympathetic to Harris's emphasis on peer, versus parental, influence, but think that our model renders moot the distinction she makes. The key issue is this: who–at any given moment–offers the best access to resources for you in your marketplace. Sometimes the answer will be your parents, sometimes your peers, sometimes your bank manager, sometimes your grandparents, sometimes your parole officer. It depends on your individual circumstances–which is precisely the message of this book. For additional background to Chapter 8, see Freud, Skinner, and Maslow and their biographers, Sulloway, Bjork, and Hoffman. Our vision of the unification of knowledge ("At One with the Universe") is paralleled by Wilson. Finally, see Holden for the textual references to Olivier; and Woolf for *Orlando.*

Allman, John Morgan. *Evolving Brains.* New York: Scientific American Library, 1999.

Barkow, Jerome H., Leda Cosmides, and John Tooby (eds.). *The Adapted Mind.* New York: Oxford University Press, 1992.

Bjork, Daniel W. *B.F. Skinner: A Life.* New York: Basic Books, 1993.

Bowlby, John. *Charles Darwin: A New Life.* New York: W.W. Norton, 1991.

Damasio, Antonio. *The Feeling of What Happens.* New York: Harcourt Brace & Company, 1999.

Darwin, Charles. *The Origin of Species.* Gillian Beer, ed. Oxford: Oxford University Press World's Classics, 1998.

——. *The Voyage of the Beagle: Charles Darwin's Journal of Researches.* Janet Browne, Michael Neve, eds. New York: Penguin USA, 1989.

Etcoff, Nancy. *Survival of the Prettiest.* New York: Doubleday, 1999.

Freud, Sigmund. *The Freud Reader.* Peter Gay, ed. New York: W.W. Norton, 1989.

Harris, Judith Rich. *The Nurture Assumption.* New York: The Free Press, 1998.

Hoffman, Edward. *The Right to Be Human: A Biography of Abraham Maslow*. Los Angeles: Tarcher, 1988.

Holden, Anthony. *Olivier*. London: Sphere Books, 1989.

James, William. *The Principles of Psychology*. New York: Henry Holt, 1890.

Maslow, Abraham. *Motivation and Personality*. Third edition. New York: HarperCollins, 1987.

Pinker, Steven. *How the Mind Works*. New York: W.W. Norton, 1997.

Ramachandran, V.S., and Sandra Blakeslee. *Phantoms in the Brain*. New York: William Morrow and Company, 1998.

Skinner, B.F. *Beyond Freedom and Dignity*. New York: Alfred Knopf, 1971.

Sulloway, Frank J. *Freud: Biologist of the Mind*. New York: Basic Books, 1979.

Weiner, Jonathan. *The Beak of the Finch*. New York: Alfred Knopf, 1994.

Wilson, Edward O. *Consilience: The Unity of Knowledge*. New York: Alfred Knopf, 1998.

Woolf, Virginia. *Orlando*. New York: Harcourt Brace & Company, n.d.

# ACKNOWLEDGMENTS

Because *The Origin of Minds* is written for a general audience, we have avoided the usual academic apparatus of footnotes and references. That said, we are very mindful of the debt we owe to those researchers whose names do not appear in the text and whose work we have used to illustrate some aspect of this new framework for understanding the origin of minds. We would like to express our gratitude to the following:

*On the bioenergetics of* E. coli: Howard Berg, Daniel Koshland.

*On the evolution of brains:* John Allman.

*On foraging behavior/mechanisms of associative learning in bees:* Leslie Real, Terrence Sejnowski, Read Montague, Peter Dayan, Martin Hammer, Randolf Menzel.

*On biological markets:* Ronald Noë, Peter Hammerstein.

*On maternal-fetal conflict:* David Haig.

*On vervet monkeys:* Michael McGuire, Michael Raleigh.

*On cichlid fish and social status:* Russell Fernald.

*On the basal ganglia and dopaminergic pathways:* Gregory Berns, Terrence Sejnowski, Peter Dayan, Read Montague, Wolfram Schultz, Trevor Robbins, Barry Everitt, Jon Horvitz.

*On neural plasticity:* Mriganka Sur, V. S. Ramachandran, Alvaro Pascual-Leone, Michael Merzenich, Charles Gross.

*On cholesterol:* Beatrice Golomb, Jay Kaplan.

*On life history evolution:* Caleb Finch, Michael Rose, George Williams.

We are very grateful for the encouragement and advice of a number of colleagues during the research period that culminated in this book, including John Allman, Patricia and Paul Churchland, Roger Masters, Michael McGuire, V. S. Ramachandran, and Terrence Sejnowski. We thank George Williams for his sage advice on the *PNAS* paper (and for communicating it); Edward O. Wilson for an appreciative response that gave us a much-needed shot in the arm; and Pat Churchland, Dan Dennett, Leslie Brothers, and Terry Deacon for thoughtful correspondence after publication.

Various drafts of the manuscript were read in whole or in part by John Allman, Linda Blitz, Deanna Clear, Michelle Grier, Jeanne Hale, Beatrice Golomb, Rosemary Kelly, V. S. Ramachandran, Terrence Sejnowski, and Martin Weiner, and we are grateful for their expenditure of energetic resources.

Our production editor, Jean Lynch, orchestrated the metamorphosis from heavily annotated manuscript pages to the book you are now reading. Matthew Budman, our copy-editor, saved us from numerous embarrassments and lapses of memory. We gladly absolve him—and the readers of the manuscript—from any remaining errors (which are, of course, ours alone).

**Peggy La Cerra expresses her sincere gratitude to—**

Ed Hagen for posing a critical research problem, and engaging me in his efforts to grapple with it back in the Fall of 1995. Ed's problem lit the fuse that imploded the Evolutionary Psychologists' model of the mind in the conceptual networks of my own, and its resultant solution laid the groundwork for the energetic evolutionary models of the mind and intelligence systems to construct themselves in its stead.

Roger Bingham for sensing the breadth of this new evolutionary framework, for researching numerous and various scientific literatures to help develop the argument that supports it, and for bringing his expansive base of general knowledge to bear on the enormously difficult task of ensuring that readers seeking to understand the universal and unique wonders of their minds now can.

J. Anthony Deutsch, Aaron Ettenberg, and Alan Fridlund for facilitating my basic research training in homeostatic, dopaminergic, and emotional systems, respectively; to Don Symons for fostering my early interest in evolutionary psychology; to Leda Cosmides and John Tooby for encouraging me to think about psychological phenomena as adaptively constrained information system processes; the McDonnell Foundation for supporting my fellowship to the McDonnell Summer Institute in Cognitive Neuroscience on Evolutionary Psychology and Neural Plasticity (where the rubber started to hit the road); and Vince Pisani, who ensured that I got on the right track in the first place.

Muriel Nellis, for her deep wisdom, sustaining kindness and unerring creative professionalism; Jane Roberts for her intuitively keen advice and vital encouragement; and everyone at Literary and Creative Artists, Inc., for their unfailing competence.

Peter Guzzardi for his superb editorial direction and guidance during the initial stages of this project, and for his support and wise counsel at a crucial juncture; Jake Morrissey for his on-the-mark editorial suggestions and refinements, and for bringing this project to fruition with seamless grace; and Kathryn Henderson, for her excellent professional assistance.

Cindy Iliff for artfully guiding the essential living elements to a viable place. It could not have happened without her.

Gail Shannon and Donald Friedman for providing me with opportunities that sustained my basic existence during the lean times.

I owe an enormous debt of gratitude to many friends, who provided me with love and support along the way—Bruce Anderson and Deborah Brown; Jaden Bennett-Andrade and his family; Kendra and Julius Braunschweig; Deanna Clear; Riki Dennis; Viola Hall; John and Anna Hench; Roxanna Javid; Darren Weiner; Sam McKinney; Allyson Rathkamp; Jason Sigel; and Rex Romero.

Special thanks to Roberto Refinetti, for vital support at a critical time; Lisa Farwell for unfailing friendship and multi-dimensional support (including serving as my expert behavioral science informant); Rosemary Kelly for being a first-rate friend, colleague and life-sustaining house mate during the sketchiest stretch of this project; Meg Miller, Dean Dawson and Owen John Dawson, who made sure they were there at the worst of times, and generously hosted the best of them; Tom Stefl, whose creative solutions to real world problems, loving kindness and generosity of spirit saved the day repeatedly; and Martin Weiner, who came at the end, and made all the difference.

Finally, thanks to my wonderful family—Gae, Fred, Marge, Michael, Laurie, Lisa, and Justin for their solid foundational support and love through the years; and Behzad Boroumand for his loving friendship, for understanding and championing the ideas in this book since their inception, for his vitally important counsel, and for being an unparalleled cooperator in the fluctuating markets of life.

**Roger Bingham additionally acknowledges:**

Over the years, I have benefited from conversations with friends and colleagues about the endlessly intriguing matter of the mind and the underpinnings of behavior—particularly John Allman (neurobiology and primate brain evolution); V.S. Ramachandran (neural plasticity and cognitive neuropsychology); Terrence Sejnowski (computational neurobiology and basal ganglia); and Michael McGuire (monoamines and primate dominance hierarchies). A special thanks to John Allman: our discussions range over two decades, which ranks him as the longest-suffering of the group. His passion for science has been an inspiration to me.

For over a decade, I have enjoyed sharing ideas—and excellent meals—with members of the Helmholtz Club, who know who they are.

For stimulating discussions about evolutionary psychology, I am grateful to Leda Cosmides, John Tooby, and—especially—

Don Symons. Our intellectual paths have since diverged, and I suspect that they will not be receptive to all the ideas in this book. Nevertheless, I would like to acknowledge the influence of their paradigm.

Peggy La Cerra insisted that there was a better–dare I say truer–model of the mind to be described. Her penetrating insights persuaded me to revise my views. The argument flew in the face of the entrenched view of evolutionary psychology (to which we had both previously subscribed).

My sister-in-law, Melody Katz, and her husband, Sam Katz, introduced me to Muriel Nellis, who became the agent for *The Origin of Minds*. Muriel and her colleague Jane Roberts (of Literary and Creative Artists, Inc.) provided support, advice, sticks and carrots, and editorial guidance, for which I am grateful.

Peter Guzzardi, our first editor, has since become an independent consultant, but I have continued to seek his advice. (If you're writing a science book, talking to the man who edited Stephen Hawking's *A Brief History of Time* is rather like asking Tiger Woods for a few pointers on your swing.) At a critical juncture for the book, Peter's wise counsel was much appreciated. His notes on later drafts were invaluable.

I was delighted that his replacement as executive editor, Jake Morrissey, was equally enthusiastic about our book and proved to be an artist with his red pencil. (Jake's first novel was published during our editorial voyage together, thus disproving the old saw that those who can't write, edit.) In fostering my understanding of the arcana of proofs, galleys, and permissions, Kathryn Henderson was a convivial guide.

There were many other people who helped me in various ways–including Lyndy Belchère, Bruce Gladstone, Jim Ivester, Sheila McHenry, Richard Payne, Steve Ridgeway, and Patrick Windmiller.

Special thanks to Cindy Iliff for her encouragement and caring diplomacy at a difficult time during the book's infancy.

And to Baruch for inspiration.

I gratefully acknowledge the L. K. Whittier Foundation for

a grant that helped make possible my contributions to the *PNAS* paper and *The Origin of Minds*.

I salute the entire Blitz family–especially Muni, Chuck, and Ronnie, who supplied the wheels–for so many kindnesses.

Thanks to Jeanne and Simon Hale for cherished moments on Cheshire retreats.

My debt to my parents is incalculable. My father, Arthur, was a man of great courage and deep curiosity and I think he would have delighted in this book. My mother, Florence, has given me a lifetime of loving encouragement, support at critical times, and–by teaching me to read–helped inspire my love of language. But my parents' greatest gift to me has been my sister, Jeanne.

Above all, I thank my wife, Linda–who knows why.

# INDEX

# ABOUT THE AUTHORS

**Peggy La Cerra** developed the scientific models of the mind and life intelligence systems presented in *The Origin of Minds* after completing an award-winning dissertation in evolutionary psychology and a McDonnell Foundation research fellowship in cognitive neuroscience. She currently serves as the director of the Center for Evolutionary Neuroscience and is the president of Mind-Works, a scientific consulting firm based in Ojai, California.

**Roger Bingham** is a member of the research faculty at the Center for Brain and Cognition at the University of California, San Diego, focusing on theoretical evolutionary neuroscience. He coauthored (with Peggy La Cerra) the paper in the *Proceedings of the National Academy of Sciences* that forms the blueprint for this new evolutionary model of the mind. Previously, he was a visiting associate in biology at the California Institute of Technology and the creator and host of award-winning PBS science programs on evolutionary psychology and cognitive neuroscience.